ISAAC ASIMOV i foremost writer on An Associate Prof the Boston Univers he has written well over a hundred books, as well as hundreds of articles in publications ranging from *Esquire* to Atomic Energy Commission pamphlets. Famed for his science fiction writing (his three-volume Hugo Award-winning THE FOUNDATION TRILOGY is available in individual Avon editions and as a one-volume Equinox edition), Dr. Asimov is equally acclaimed for such standards of science reportage as THE UNIVERSE, LIFE AND ENERGY, THE SOLAR SYSTEM AND BACK, ASIMOV'S BIOGRAPHICAL ENCYCLOPEDIA OF SCIENCE AND TECHNOLOGY, and ADDING A DIMENSION (all available in Avon editions). His non-science writings include the two-volume ASIMOV'S GUIDE TO SHAKESPEARE, ASIMOV'S ANNOTATED DON JUAN, and the two-volume ASIMOV'S GUIDE TO THE BIBLE (available in a two-volume Avon edition). Born in Russia, Asimov came to this country with his parents at the age of three, and grew up in Brooklyn. In 1948 he received his Ph.D. in Chemistry at Columbia and then joined the faculty at Boston University, where he works today.

Avon Books by
Isaac Asimov

VIEW FROM
A HEIGHT

ISAAC ASIMOV

 DISCUS BOOKS/PUBLISHED BY AVON

AVON BOOKS
A division of
The Hearst Corporation
959 Eighth Avenue
New York, New York 10019

ISBN: 0-380-00356-2

First Discus Printing, June, 1975.

DISCUS TRADEMARK REG. U.S. PAT. OFF. AND
FOREIGN COUNTRIES, REGISTERED TRADEMARK—
MARCA REGISTRADA, HECHO EN CHICAGO, U.S.A.

Printed in the U.S.A.

Contents

Introduction

Science, before 1800, was like an orchard: tame, well laid-out and ordered; fragrant and fruit-bearing. One could wander through it from end to end observing its various parts; and from a neighboring hill, one might even see the entire scheme, and appreciate it.

But by 1800, the wanderers noted that the busy planters, gardeners and cultivators had done their work so well that parts of it had grown dark and foreboding. Order was still there, indeed the intricate network of relationships was more refined, more subtle and more fascinating than ever; but the proliferating branches had begun to shut out the sky.

And then there came the shock of realizing that the orchard was too large. One could no longer pass through it from end to end—not without getting lost and walking in circles back to one's starting point. Nor was the neighboring hill any longer of use, for it, too, was covered by orchard now.

So some of the observers, some of the lovers of the beauty of order, abandoned the orchard altogether; while others compromised by confining themselves to narrow sections of the orchard, then to narrower sections and still narrower sections.

Now, the orchard of science is a vast globe-encircling monster, without a map, and known to no one man; indeed, to no group of men fewer than the whole international mass of creative scientists. Within it, each observer clings to his own well-known and well-loved clump of trees. If he looks beyond, it is usually with a guilty sigh.

And just as an organism in the embryonic stages seems to race through the aeons of evolutionary development, from the single cell to ultimate complexity, in mere weeks or months; so the individual scientist in the course of his life repeats the history of science and loses himself by progressive stages, in the orchard.

When I was young, the neighborhood public libraries were my source material. I lacked the intelligence and purpose to be selective, so I started at one end of the shelves and worked my way toward the other.

Since I have an inconveniently retentive memory, I learned a great many things in this way that I have since labored unsuccessfully to forget. One of the valuable things I learned, however, was that I liked nonfiction better than fiction; history better than most other varieties of nonfiction; and science even better than history.

By the time I entered high school, I had restricted myself largely to history and science. By the time I entered college, I was down to science.

In college, I found I was expected to choose among the major disciplines of science and select a "major." I flirted with zoology and then, in my second year, made chemistry a firm choice. This meant not much more than one chemistry course per semester, but when I entered graduate school, I found that all my courses were chemistry of one sort or another, and I had to choose among them with a view toward doctoral research.

Through a series of developments of absorbing lack of interest (as far as these pages are concerned), I found myself doing research on a biochemical topic. In that area of study I obtained my Ph.D., and in no time at all I was teaching biochemistry at a medical school.

But even that was too wide a subject. From books to nonfiction, to science, to chemistry, to biochemistry—and not yet enough. The orchard had to be narrowed down further. To do research, I had to find myself a niche within biochemistry, so I began work on nucleic acids. . . .

And at about that point, I rebelled! I could not stand the claustrophobia that clamped down upon me. I looked with horror, backward and forward across the years, at a horizon that was narrowing down and narrowing down to so petty a portion of the orchard. What I wanted was all the orchard, or as much of it as I could cover in a life-time of running.

To be sure, rebellion at this stage is often useless. Specialization has taken hold firmly and one dare not leave its confines. The faculty of participating in life outside one's special field has atrophied.

But fortunately, through all my higher education from college on, I had stubbornly maintained a status as an active science fiction writer (partly because I needed the money, but mostly because I loved it). To be a proper science fiction writer (in my opinion) one must somehow retain a nodding acquaintance with as many of the branches of science as possible; and this, of course, I had tried to do.

So when I decided to push away from specialization, my science fiction afforded me two saving assets. First, I had a source of income in case anything went wrong. Second, I had never lost a toe hold in other parts of the orchard.

I have never been sorry for my stubborn advance toward generalization. To be sure, I can't wander in detail through all the orchard, any more than anyone else can, no matter how stupidly determined I may be to do so. Life is far too short and the mind is far too limited. But I can float over the orchard as in a balloon.

I see no signs of an end; I make out very little detail. The precious intricacy one can observe by crawling over half an acre is lost to me.

But I have gains of my own, for occasionally I see (or seem to see) an overall order, or an odd arabesque in one corner—a glimmering fragment of design that perhaps I would not see from the ground.

And when I do (or think I do), I set it down on paper, for in addition to my other peculiarities, I have a mission-ary complex and want others to see what I see. Fortunately, I have a way with editors, and can harry them into publish-ing what I set down.

So I have here a collection of essays with little internal

unity. They are glimpses, here and there, of the orchard of science, as viewed from a height.

And the reason for the collection is only that I do want, very much, to have you see what I see.

I BIOLOGY

1 That's About the Size of It

No matter how much we tell ourselves that quality is what counts, sheer size remains impressive. The two most popular types of animals in any zoo are the monkeys and the elephants, the former because they are embarrassingly like ourselves, the latter simply because they are huge. We laugh at the monkeys but stand in silent awe before the elephant. And even among the monkeys, if one were to place Gargantua in a cage, he would outdraw every other primate in the place. In fact, he did.

This emphasis on the huge naturally makes the human being feel small, even puny. The fact that mankind has nevertheless reached a position of unparalleled domination of the planet is consequently presented very often as a David-and-Goliath saga, with ourselves as David.

And yet this picture of ourselves is not quite accurate, as we can see if we view the statistics properly.

First, let's consider the upper portion of the scale. I've just mentioned the elephant as an example of great size, and this is hallowed by cliché. "Big as an elephant" is a common phrase.

But, of course, the elephant does not set an unqualified

record. No land animal can be expected to. On land, an animal must fight gravity, undiluted. Even if it were not a question of lifting its bulk several feet off the ground and moving it more or less rapidly, that fight sets sharp limits to size. If an animal were envisaged as lying flat on the ground, and living out its life as motionlessly as an oyster, it would still have to heave masses of tissue upward with every breath. A beached whale dies for several reasons, but one is that its own weight upon its lungs slowly strangles it to death.

In the water, however, buoyancy largely negates gravity, and a mass that would mean crushing death on land is supported under water without trouble.

For that reason, the largest creatures on earth, present or past, are to be found among the whales. And the species of whale that holds the record is the blue whale, or, as it is alternatively known, the sulfur-bottom. One specimen of this greatest of giants has been recorded with a length of 108 feet and a weight of 131¼ tons.

Now the blue whale, like ourselves, is a mammal. If we want to see how we stand among the mammals, as far as size is concerned, let's see what the other extreme is like.

The smallest mammals are the shrews, creatures that look superficially mouselike, but are not mice or even rodents. Rather, they are insectivores, and are actually more closely related to us than to mice. The smallest full-grown shrew weighs about one-thirteenth of an ounce.

Between these two mammalian extremes stretch a solid phalanx of animals. Below the blue whale are other smaller whales, then creatures such as elephants, walruses, hippopotamuses, down through moose, bears, bison, horses, lions, wolves, beavers, rabbits, rats, mice and shrews. Where in this long list from largest whale to smallest shrew is man?

To avoid any complications, and partly because my weight comes to a good, round figure of two hundred pounds, I will use myself as a measure.

Now, we can consider man either a giant or a pygmy, according to the frame of reference. Compared to the shrew he is a giant, of course, and compared to the whale he is a pygmy. How do we decide which view to give the greater weight?

In the first place, it is confusing to compare tons, pounds and ounces, so let's put all three weights into a common unit. In order to avoid fractions (just at first, anyway) let's consider grams as the common unit. (For your reference, one ounce equals about 28.35 grams, one pound equals about 453.6 grams, and one ton equals about 907,000 grams.)

Now, you see, we can say that a blue whale weighs as much as 120,000,000 grams while a shrew weighs as little as 2 grams. In between is man with a weight of 90,700 grams.

We are tens of thousands of grams heavier than a shrew, but a whale is tens of *millions* of grams heavier than a man, so we might insist that we are much more of a pygmy than a giant and insist on retaining the David-and-Goliath picture.

But human sense and judgment do not differentiate by subtraction; they do so by division. The differences between a two-pound weight and a six-pound weight seems greater to us than that between a six-pound weight and a twelve-pound weight, even though the difference is only four pounds in the first case and fully six pounds in the latter. What counts, it seems, is that six divided by two is three, while twelve divided by six is only two. Ratio, not difference, is what we are after.

Naturally, it is tedious to divide. As any fourth-grader and many adults will maintain, division comes under the heading of advanced mathematics. Therefore, it would be pleasant if we could obtain ratios by subtraction.

To do this, we take the logarithm of a number, rather than the number itself. For instance, the most common form of logarithms are set up in such a fashion that 1 is the logarithm of 10, 2 is the logarithm of 100, 3 is the logarithm of 1,000 and so on.

If we use the numbers themselves, we would point out an equality of ratio by saying that 1,000/100 is equal to 100/10, which is division. But if we used the logarithms, we could point out the same quality of ratio by saying that 3 minus 2 is equal to 2 minus 1, which is subtraction.

Or, again, 1000/316 is roughly equal to 316/100. (Check it and see.) Since the logarithm of 1,000 is 3 and the logarithm of 100 is 2, we can set the logarithm of 316 equal to 2.5, and then, using logarithms, we can express the equality of ratio by saying that 3 minus 2.5 is equal to 2.5 minus 2.

So let's give the extremes of mammalian weight in terms of the logarithm of the number of grams. The 120,000,000-gram blue whale can be represented logarithmically by 8.08, while the 2-gram shrew is 0.30. As for the 90,700-gram man, he is 4.96.

As you see, man is about 4.7 logarithmic units removed from the shrew but only about 3.1 logarithmic units removed from the largest whale. We are therefore much more nearly giants than pygmies.

In case you think all this is mathematical folderol and that I am pulling a fast one, what I'm saying is merely equivalent to this: A man is 45,000 times as massive as a shrew, but a blue whale is only 1,300 times as massive as a man. We would seem much larger to a shrew than a whale does to us.

In fact, a mass that would be just intermediate between that of a shrew and a whale would be one with a logarithm that is the arithmetical average of 0.30 and 8.08, or 4.19. This logarithm represents a mass of 15,550 grams, or 34 pounds. By that argument a medium-sized mammal would be about the size of a four-year-old child, or a dog of moderate weight.

Of course, you might argue that a division into two groups—pygmy and giant—is too simple. Why not a division into three groups—pygmy, moderate, and giant? Splitting the logarithmic range into three equal parts, we would have the pygmies in the range from 0.30 to 2.90, the moderates from 2.90 to 5.40, and the giants from 5.40 to 8.08.

Put into common units this would mean that any animal under 1¾ pounds would be a pygmy and any animal over 550 pounds would be a giant. By that line of thinking, the animals between, including man, would be a moderate size. This seems reasonable enough, I must admit, and it seems a fair way of showing that man, if not a pygmy, is also not a giant.

But if we're going to be fair, let's be fair all the way. The David-and-Goliath theme is introduced with respect to man's winning of overlordship on this planet; it is the victory of brains over brawn. Early man never competed with whales. Whales stayed in the ocean and man stayed on land. Our battle was with land creatures only, so let's consider land mammals in setting up our upper limit.

The largest land mammal that ever existed is not alive today. It is the baluchitherium, an extinct giant rhinoceros that stood eighteen feet tall at the shoulder, and must have weighed in the neighborhood of 13½ tons.

As you see, the baluchitherium (which means "Baluchi beast," by the way, because its fossils were first found in Baluchistan) has only one-tenth the mass of a blue whale. Since the ratio 10 is represented by the logarithm 1, you won't be surprised to hear that the logarithmic value of the baluchitherium's mass (in grams) is 1 less than that of the blue whale and stands at 7.08.

(From now on, I will give weights in common units but will follow it with a logarithmic value in parentheses. Please remember that this is the logarithm of weight in grams every time.)

But, of course, the baluchitherium was extinct before the coming of man and there was no competition with him either. To make it reasonably fair, we must compare man with those creatures that were alive in his time and therefore represented potential competition. The largest mammals living in the time of man are the various elephants. The largest living African elephant may reach a total weight of seven tons (6.80). To be sure, it is possible that man competed with still larger species of elephant now extinct, but that makes little difference. The largest elephant that ever existed could not have weighed more than ten tons (6.96).

(Notice, by the way, that an elephant is only about half as heavy as a baluchitherium and has only 5 percent of the weight of a blue whale. In fact, a full-grown elephant of the largest living kind is only about the weight of a new-born blue whale.)

Nor am I through. In battling other species for world domination, the direct competitors to man were other carnivores. An elephant is herbivorous. It might crush a man to death accidentally, or on purpose if angered, but otherwise it has no reason to harm man. A man does not represent food to an elephant.

A man does represent food to a saber-toothed tiger, however, which, if hungry enough, would stalk, kill and eat a man who was only trying to stay out of the way. *There* is the competition.

Now the very largest animals are almost invariably herbivores. There are more plant calories available than animal calories and a vegetable diet can, on the whole, support larger animals than a meat diet. (Which is not to say that some carnivores aren't much larger than some herbivores.)

To be sure, the largest animal of all, the blue whale, is technically a carnivore. However, he lives on tiny creatures strained out of ocean water, and this isn't so far removed, in a philosophical sense, from browsing on grass. He is not a carnivore of the classic type, the kind with teeth that go snap!

The largest true carnivore in all the history of the earth is the sperm whale (of which Moby Dick is an example). A mature sperm whale, with a large mouth and a handsome set of teeth in its lower jaw, may weigh sixty tons (7.74).

But there again, we are not competing with sea creatures. The largest *land* carnivore among the mammals is the great Alaskan bear (also called the Kodiak bear), which occasionally tips the scale at 1,600 pounds (5.86). I don't know of any extinct land carnivore among the mammals that was larger.

Turning to the bottom end of the scale, there we need make no adjustments. The shrew is a land mammal and a carnivore and, as far as I know, is the smallest mammal that can possibly exist. The metabolic rate of mammals goes up as size decreases because the surface-to-volume ratio goes up with decreasing size. Some small animals might (and do) make up for that by letting the metabolic rate drop and the devil with it, but a warm-blooded creature cannot. It must keep its temperature high, and therefore, its metabolism racing (except during temporary hibernations).

A warm-blooded animal the size of a shrew must eat just about constantly to keep going. A shrew starves to death if it goes a couple of hours without eating; it is always hungry and is very vicious and ill-tempered in consequence. No one has ever seen a fat shrew or ever will. (And if anyone wishes to send pictures of the neighbor's wife in order to refute that statement, please don't.)

Now let's take the range of land-living mammalian carnivores and break that into three parts. From 0.30 to 2.15 are the pygmies, from 2.15 to 4.00 the moderates, and from 4.00 to 5.86 the giants. In common units that would mean that any creature under 5 ounces is a pygmy, anything from 5 ounces to 22 pounds is a moderate, and anything over 22 pounds is a giant.

Among the mammalian land carnivores of the era in which man struggled through first to survival and then to victory, man is a giant. In the David-and-Goliath struggle, one of the Goliaths won.

Of course, some suspicion may be aroused by the fact that I am so carefully specifying mammals throughout. Maybe man is only a giant among mammals, you may think, but were I to broaden the horizon he would turn out to be a pygmy after all.

Well, not so. As a matter of fact, mammals in general are giants among animals. Only one kind of non-mammal can compete (on land) with the large mammals, and they are the reptile monsters of the Mesozoic era—the large group of animals usually referred to in common speech as "the dinosaurs."

The largest dinosaurs were almost the length of the very largest whales, but they were mostly thin neck and thin tail, so that they cannot match those same whales in mass. The bulkiest of the large dinosaurs, the Brachiosaurus, probably weighed no more than fifty tons (7.65) at most.

This is respectable, to be sure. It is seven times the size of the baluchitherium, but it is only two-fifths the size of the blue whale. And, as is to be expected, the largest of the dinosaurs were herbivorous.

The largest carnivorous dinosaur was the famous Tyrannosaurus rex, which weighed perhaps as much as fifteen tons (7.13). It is clearly larger than the baluchitherium, rather more than twice the weight of the African elephant and nearly twenty times the weight of the poor little Kodiak bear.

The Tyrannosaurus rex was, beyond doubt, the largest and most fearsome land carnivore that ever lived. He and all his tribe, however, were gone from the earth millions of years before man appeared on the scene.

If we confine ourselves to reptiles alive in man's time, the largest appear to be certain giant crocodiles of Southeast Asia. Unfortunately, reports about the size of such creatures always tend to concentrate on the length rather than the weight (this is even truer of snakes); some are described as approaching thirty feet in length. I estimate that such monsters should also approach a maximum of two tons (6.25) in weight.

I have a more precise figure for the next most massive group of living reptiles, the turtles. The largest turtle on record is a marine leatherback with a weight of 1,902 pounds (5.93), or not quite a ton.

To be sure, neither of these creatures are land animals. The leatherback is definitely a creature of the sea, while crocodiles are river creatures. Nevertheless, as far as the crocodiles are concerned I am inclined not to omit them from the list of man's competitors. Early civilizations developed along tropical or subtropical rivers; who is not aware of the menace of the crocodile of the Nile, for instance? And certainly it is a dangerous creature with a mouth and teeth that go snap! to end all snaps! (What jungle movie would omit the terrifying glide and gape of the crocodile?)

The crocodiles are smaller than the largest land-living mammals, but the largest of these reptiles would seem to outweigh the Kodiak bear. However, even if we let 5.93 be the new upper limit of the "land" carnivores, man would still count as a giant.

If we move to reptiles that are truly of the land, their

inferiority to mammals in point of size is clear. The largest land reptile is the Galapagos tortoise, which may reach five hundred pounds (5.35). The largest snake is the reticulated python, which may reach an extreme length of thirty-three feet. Here again, weights aren't given, as all the ooh'ing and ah'ing are over the measurement by yardstick. However, I don't see how this can represent a weight greater than 450 pounds (5.32). Finally, the largest living lizard is the Komodo monitor, which grows to a maximum length of twelve feet and to a weight of 250 pounds (5.05).

The fishes make a fairly respectable showing. The largest of all fishes, living or extinct, is the whale shark. The largest specimens of these are supposed to be as large and as massive as the sperm whale, though perhaps a forty-five-ton maximum (7.61) might be more realistic. Again, these sharks are harmless filterers of sea water. The largest carnivorous shark is the white shark, which reaches lengths of twenty feet and possibly a weight of two and a half tons (6.36).

Of the bony fishes, the largest (such as the tuna, swordfish or sturgeon) may tip the scales at as much as three thousand pounds (6.13). All fish, however, are water creatures, of course, and not direct competition for any man not engaged in such highly specialized occupations as pearl-diving.

The birds, as you might expect, make a poorer showing. You can't be very heavy and still fly.

This means that any bird that competes with man in weight must be flightless. The heaviest bird that ever lived was the flightless Aepyornis of Madagascar (also called the elephant bird), which stood ten feet high and may have weighed as much as one thousand pounds (5.66). The largest moas of New Zealand were even taller (twelve feet) but were more lightly built and did not weigh more than five hundred pounds (5.36). In comparison, the largest living bird, the ostrich—still flightless—has a maximum weight of about three hundred pounds (5.13).

When we get to flying birds, weight drops drastically. The albatross has a record wingspread of twelve feet, but wings don't weigh much and even the heaviest flying bird probably does not weigh more than thirty-five pounds

(4.20). Even the pteranodon, which was the largest of the extinct flying reptiles, and had a wingspread of up to twenty-five feet, was virtually all wing and no body, and probably weighed less than an albatross.

To complete the classes of the vertebrates, the largest amphibians are giant salamanders found in Japan, which are up to five feet in length and weigh up to ninety pounds (4.60).

Working in the other direction, we find that the smallest bird, the bee hummingbird of Cuba, is probably about the size of the smallest shrew. (Hummingbirds also have to keep eating almost all the time, and starve quickly.)

The cold-blooded vertebrates can manage smaller sizes than any of the warm-blooded mammals and birds, however, since cold blood implies that body temperature can drop to that of the surroundings and metabolism can be lowered to practical levels. The smallest vertebrates of all are therefore certain species of fish. There is a fish of the goby group in the Philippine Islands that has a length, when full grown, of only three-eighths of an inch. I estimate its weight would be no more than 0.3 grams (-0.52), which, as you notice, carries us into negative logarithms.

What about invertebrates?

Well, invertebrates, having no internal skeleton with which to brace their tissues, cannot be expected to grow as large as vertebrates. Only in the water, where they can count on buoyancy, can they make any decent showing at all.

The largest invertebrates of all are to be found among the mollusks. Giant squids with lengths up to fifty-five feet have been actually measured, and lengths up to one hundred feet have been conjectured. Even so, such lengths are illusory, for they include the relatively light tentacles for the most part. The total weight of such creatures is not likely to be much more than two tons (6.26).

Another type of mollusk, the giant clam, may reach a weight of seven hundred pounds (5.50), mostly dead shell, while the largest arthropod is a lobster that weighed in at thirty-four pounds (4.19).

As for the land invertebrates, mass is negligible. The

largest land crabs and land snails never match the weights of any but quite small mammals. The same is true of the most successful and important of all the land invertebrates, the insects. The bulkiest insect is the goliath beetle, which can be up to four or five inches in length. I can find no record of what it weighs, but I should judge that weight to be in the neighborhood of an ounce (1.44).

And the insects, with a top weight just overlapping the bottom of the mammalian scale, are well represented in levels of less and less massive creatures. The bottom is an astonishing one, for there are small beetles called fairy flies that are as small as $\frac{1}{125}$ of an inch in length, full-grown. Such creatures can have weights of no more than 0.0000001 grams (−7.00).

Nor is even this the record. Among the various classes of multicelled invertebrates, the smallest of all is Rotifera. Even the largest of these are only one-fifteenth of an inch long, while the smallest are but one three-hundredth of an inch long and may weigh 0.000000006 grams (−8.22). The rotifers, in other words, are to the shrews as the shrews are to the whales. If we go still lower, we will end considering not only man but also the shrew as a giant among living creatures.

But below the rotifers are the one-celled creatures (though, in fact, the larger one-celled creatures are larger than the smallest rotifers and insects), and I will stop here, adding only a summarizing table of sizes.

But if we are to go back to the picture of David and Goliath, and consider man a Goliath, we have some real Davids to consider—rodents, insects, bacteria, viruses. Come to think of it, the returns aren't yet in, and the wise money might be on the real Davids after all.

TABLE OF SIZES

Animal	Characteristic	Logarithm of Weight in Grams
Blue whale	Largest of all animals	8.08
Sperm whale	Largest of all carnivores	7.74

Brachiosaurus	Largest land animal (extinct)	7.65
Whale shark	Largest fish	7.61
Tyrannosaurus rex	Largest land carnivore (extinct)	7.13
Baluchitherium	Largest land mammal (extinct)	7.08
Elephant	Largest land animal (alive)	6.80
White shark	Largest carnivorous fish	6.36
Giant squid	Largest invertebrate	6.26
Crocodile	Largest reptile (alive)	6.26
Sturgeon	Largest bony fish	6.13
Leatherback	Largest turtle	5.93
Kodiak bear	Largest land carnivore (alive)	5.86
Aepyornis	Largest bird (extinct)	5.66
Giant clam	Largest gastropod	5.50
Galapagos tortoise	Largest land reptile (alive)	5.35
Reticulated python	Largest snake	5.32
Ostrich	Largest bird (alive)	5.13
Komodo monitor	Largest lizard	5.05
Man		4.96
Giant salamander	Largest amphibian	4.60
Albatross	Largest flying bird	4.20
Lobster	Largest arthropod	4.20
Goliath beetle	Largest insect	1.44
Shrew	Smallest mammal	0.30
Bee hummingbird	Smallest bird	0.30
Goby (fish)	Smallest vertebrate	−0.52
Fairy fly	Smallest insect	−7.00
Rotifer	Smallest multicelled creature	−8.22

2 The Egg and Wee

Every once in a while, you will come across some remarks pointing up how much more compact the human brain is than is any electronic computer.

It is true that the human brain is a marvel of compactness in comparison to man-made thinking machines, but it is my feeling that this is not because of any fundamental difference in the nature of the mechanism of brain action as compared with that of computer action. Rather, I have the feeling that the difference is a matter of the size of the components involved.

The human cerebral cortex, it is estimated, is made up of 10,000,000,000 nerve cells. In comparison, the first modern electronic computer, ENIAC, had about 20,000 switching devices. I don't know how many the latest computers have, but I am quite certain they do not begin to approach a content of ten billion.

The marvel, then, is not so much the brain as the cell. Not only is the cell considerably smaller than any man-made unit incorporated into a machine, but it is far more flexible than any man-made unit. In addition to acting as an electronic switch or amplifier (or whatever it does in the brain), it is a complete chemical factory.

Furthermore, cells need not aggregate in fearfully large numbers in order to make up an organism. To be sure, the average man may contain 50,000,000,000,000 (fifty trillion) cells and the largest whale as many as 100,000,000,-000,000,000 (a hundred quadrillion) cells, but these are exceptional. The smallest shrew contains only 7,000,000,-000 cells, and small invertebrate creatures contain even less. The smallest invertebrates are made up of only one hundred cells or so, and yet fulfill all the functions of a living organism.

As a matter of fact (and I'm sure you're ahead of me here), there are living organisms that possess all the basic abilities of life and are nevertheless composed of but a single cell.

If we are going to concern ourselves with compactness, then, let's consider the cell and ask ourselves the questions: How compact can a living structure be? How small can an object be and still have the capacity for life?

To begin: How large is a cell?

There is no one answer to that, for there are cells and cells, and some are larger than others. Almost all are microscopic, but some are so large as to be clearly, and even unavoidably, visible to the unaided eye. Just to push it to an extreme, it is possible for a cell to be larger than your head.

The giants of the cellular world are the various egg cells produced by animals. The human egg cell (or ovum), for instance, is the largest cell produced by the human body (either sex), and it is just visible to the naked eye. It is about the size of a pinhead.

In order to make the size quantitative and compare the human ovum in reasonable fashion with other cells both larger and smaller, let's pick out a convenient measuring unit. The inch or even the millimeter (which is approximately $\frac{1}{25.4}$ of an inch) is too large a unit for any cell except certain egg cells. Instead, therefore, I'm going to use the micron, which equals a thousandth of a millimeter or $\frac{1}{25,400}$ of an inch. For volume, we will use a cubic micron, which is the volume of a cube one micron long on each side. This is a very tiny unit of volume, as you will understand when I tell you that a cubic inch (which is something that is easy to visualize) contains over 16,000,-000,000,000 (sixteenth trillion) cubic micra.

There are a third as many cubic micra in a cubic inch, then, as there are cells in a human body. That alone should tell us we have a unit of the right magnitude to handle cellular volumes.

24

Back to the egg cells then. The human ovum is a little sphere approximately 140 micra in diameter and therefore 70 micra in radius. Cubing 70 and multiplying the result by 4.18 (I will spare you both the rationale and the details of arithmetic manipulation), we find that the human ovum has a volume of a little over 1,400,000 cubic micra.

But the human ovum is by no means large for an egg cell. Creatures that lay eggs, birds in particular, do much better; and bird eggs, however large, are (to begin with, at least) single cells.

The largest egg ever laid by any bird was that of the extinct Aepyornis of Madagascar. This was also called the elephant bird, and may have given rise to the myth—so it is said—of the roc of the *Arabian Nights*. The roc was supposed to be so large that it could fly off with an elephant in one set of talons and a rhinoceros in the other. Its egg was the size of a house.

Actually, the Aepyornis was not quite that lyrically vast. It could not fly off with any animal, however small, for it could not fly at all. And its egg was considerably less than house-size. Nevertheless, the egg was nine and one-half inches wide and thirteen inches long and had a volume of two gallons, which is tremendous enough if you want to restrict yourself to the dullness of reality.

This is not only the largest egg ever laid by any bird, but it may be the largest ever laid by any creature, including the huge reptiles of the Mesozoic age, for the Aepyornis egg approached the maximum size that any egg, with a shell of calcium carbonate and without any internal struts or braces, can be expected to reach. If the Aepyornis egg is accepted as the largest egg, then it is also the largest cell of which there is any record.

To return to the here and now, the largest egg (and, therefore, cell) produced by any living creature is that of the ostrich. This is about six to seven inches in length and four to six inches in diameter; and, if you are interested, it takes forty minutes to hard-boil an ostrich egg. In comparison, a large hen's egg is about one and three-quarter inches wide and two and a half inches long. The smallest egg laid by a bird is that of species of hummingbird which produces an egg that is half an inch long.

25

Now let's put these figures, very roughly, into terms of volume:

Egg	Volume (in cubic micra)
Aepyornis	7,500,000,000,000,000
Ostrich	1,100,000,000,000,000
Hen	50,000,000,000,000
Hummingbird	400,000,000,000
Human being	1,400,000

As you see, the range in the egg size is tremendous. Even the smallest bird egg is about 300,000 times as voluminous as the human ovum, whereas the largest bird egg is nearly 20,000 times as large as the smallest.

In other words, the Aepyornis egg compares to the hummingbird egg as the largest whale compares to a medium-sized dog; while the hummingbird egg, in turn, compares to the human ovum as the largest whale compares to a large rat.

And yet, even though the egg consists of but one cell, it is not the kind of cell we can consider typical. For one thing, scarcely any of it is alive. The eggshell certainly isn't alive and the white of the egg serves only as a water store. The yolk of the egg makes up the true cell and even that is almost entirely food supply.

If we really want to consider the size of cells, let's tackle those that contain a food supply only large enough to last them from day to day—cells that are largely protoplasm, in other words. These non-yolk cells range from the limits of visibility upward.

In fact, there is some overlapping. For instance, the amoeba, a simple free-living organism consisting of a single cell, has a diameter of about two hundred micra and a volume of 4,200,000 cubic micra. It is three times as voluminous as the human ovum.

The cells that make up multicellular organisms are considerably smaller, however. The various cells of the human body have volumes varying from 200 to 15,000 cubic micra. A typical liver cell, for instance, would have a volume of 1,750 cubic micra.

If we include cell-like bodies that are not quite complete

cells, then we can reach smaller volumes. For instance, the human red blood cell, which is incomplete in that it lacks a cell nucleus, is considerably smaller than the ordinary cells of the human body. It has a volume of only 90 cubic micra.

Then, just as the female ovum is the largest cell produced by human beings, the male spermatozoon is the smallest. The spermatozoon is mainly nucleus, and only half the nucleus at that. It has a volume of about 17 cubic micra.

This may make it seem to you that the cells making up a multicellular organism are simply too small to be individual and independent fragments of life, and that in order to be free-living a cell must be unusually large. After all, an amoeba is 2,400 times as large as a liver cell, so perhaps in going from amoeba to liver cell, we have passed the limit of compactness that can be associated with independent life.

This is not so, however. Human cells cannot, to be sure, serve as individual organisms, but that is only because they are too specialized and *not* because they are too small. There are cells that serve as independent organisms that are far smaller than the amoeba and even smaller than the human spermatozoon. These are the bacteria.

Even the largest bacterium has a volume of no more than 7 cubic micra, while the smallest have volumes down to 0.02 cubic micra. All this can be summarized as follows:

Non-yolk cell	Volume (in cubic micra)
Amoeba	4,200,000
Human liver cell	1,750
Human red blood cell	90
Human spermatozoon	17
Largest bacterium	7
Smallest bacterium	0.02

Again we have quite a range. A large one-celled organism such as the amoeba is to a small one-celled organism such as a midget bacterium, as the largest full-grown whale is to a half-grown specimen of the smallest variety of shrew. For that matter, the difference between the largest and smallest bacterium is that between a large elephant and a small boy.

Now, then, how on earth can the complexity of life be crammed into a tiny bacterium one two-hundred-millionth the size of a simple amoeba?

Again we are faced with a problem in compactness and we must pause to consider units. When we thought of a brain in terms of pounds, it was a small bit of tissue. When we thought of it in terms of cells, however, it became a tremendously complex assemblage of small units. In the same way, in considering cells, let's stop thinking in terms of cubic micra and start considering atoms and molecules.

A cubic micron of protoplasm contains about 40,000,-000,000 molecules. Allowing for this, we can recast the last table in molecular terms:

Cell	Number of molecules
Amoeba	170,000,000,000,000,000
Human liver cell	70,000,000,000,000
Human red blood cell	3,600,000,000,000
Human spermatozoon	680,000,000,000
Largest bacterium	280,000,000,000
Smallest bacterium	800,000,000

It would be tempting, at this point, to say that the molecule is the unit of the cell, as the cell is the unit of a multicellular organism. If we say that, we can go on to maintain that the amoeba is seventeen million times as complicated, molecularly speaking, as the human brain is, cellularly speaking. In that case, the compactness of the amoeba as a container for life becomes less surprising.

There is a catch, though. Almost all the molecules in protoplasm are water; simple little H_2O combinations. These are essential to life, goodness knows, but they serve molecules of life. If we can point to any molecules as largely as background, They are not *the* characteristic molecules of life. If we can point to any molecules as characteristic of life, they are the complex nitrogen-phos-phorus macromolecules: the proteins, the nucleic acids

28

and the phospholipids. These, together, make up only about one ten-thousandth of the molecules in living tissue.

(Now, I am *not* saying that these macromolecules make up only 1/10,000 of the *weight* of living tissue; only of the numbers of molecules. The macromolecules are individually much heavier than the water molecules. An average protein molecule, for instance, is some two thousand times as heavy as a water molecule. If a system consisted of two thousand water molecules and one average protein molecule, the *number* of protein molecules would only be 1/2,001 of the total, but the *weight* of protein would be 1/2 the total.)

Let's revise the table again, then:

Cell	Nitrogen-Phosphorus Macromolecules
Amoeba	170,000,000,000,000
Human liver cell	7,000,000,000
Human red blood cell	360,000,000
Human spermatozoon	68,000,000
Largest bacterium	28,000,000
Smallest bacterium	80,000

We can say, then, that the average human body cell is indeed as complex, molecularly speaking, as the human brain, cellularly speaking. Bacteria, however, are markedly simpler than the brain, while the amoeba is markedly more complex.

Still, even the simplest bacterium grows and divides with great alacrity and there is nothing simple, from the chemical standpoint, about growing and dividing. That simplest bacterium, just visible under a good optical microscope, is a busy, self-contained and complex chemical laboratory.

But then, most of the 80,000 macromolecules in the smallest bacterium (say 50,000 at a guess) are enzymes, each of which can catalyze a particular chemical reaction. If there are 2,000 different chemical reactions constantly proceeding within a cell, each of which is necessary to growth and multiplication (this is another guess), then there are, on the average, 25 enzymes for each reaction.

A human factory in which 2,000 different machine

operations are being conducted, with 25 men on each machine, would rightly be considered a most complex structure. Even the smallest bacterium is that complex.

We can approach this from another angle, too. About the turn of the century, biochemists began to realize that in addition to the obvious atomic components of living tissue (such as carbon, hydrogen, oxygen, nitrogen, sulfur, phosphorus and so on) certain metals were required by the body in very small quantities.

As an example, consider the two most recent additions to the list of trace metals in the body, molybdenum and cobalt. The entire human body contains perhaps 18 milligrams of molybdenum and 12 milligrams of cobalt (roughly one two-thousandth of an ounce of each). Nevertheless, this quantity, while small, is absolutely essential. The body cannot exist without it.

To make this even more remarkable, the various trace minerals, including molybdenum and cobalt, seem to be essential to every cell. Divide up one two-thousandth of an ounce of these materials among the fifty trillion cells of the human body and what a miserably small trace of a trace is supplied each! *Surely,* the cells can do without.

But that is only if we persist in thinking in terms of ordinary weight units instead of in atoms. In the average cell, there are, very roughly speaking, some 40 molybdenum and cobalt atoms for every billion molecules. Let's, therefore, make still another table:

Cell	Number of molybdenum and cobalt atoms
Amoeba	6,800,000,000
Human liver cell	2,800,000
Human red blood cell	144,000
Human spermatozoon	27,200
Largest bacterium	11,200
Smallest bacterium	32

(Mind you, the cells listed are not necessarily "average." I am quite certain that the liver cell contains more than an average share of these atoms and the red blood cell less than an average share; just as earlier, the spermatozoon undoubtedly contained more than an average share of macromolecules. However, I firmly refuse to quibble.)

As you see, the trace minerals are not so sparse after all. An amoeba possesses them by the billions of atoms and a human body cell by the millions. Even the larger bacteria possess them by the thousands.

The smallest bacteria, however, have only a couple of dozen of them, and this fits in well with my earlier conclusion that the tiniest bacterium may have, on the average, 25 enzymes for each reaction. Cobalt and molybdenum (and the other trace metals) are essential because they are key bits of important enzymes. Allowing one atom per enzyme molecule, there are only a couple of dozen such molecules, all told, in the smallest bacterium.

But here we can sense that we are approaching a lower limit. The number of different enzymes is not likely to be distributed with perfect evenness. There will be more than a couple of dozen in some cases and less than a couple of dozen in others. Only one or two of the rarest of certain key enzymes may be present. If a cell had a volume of less than 0.02 cubic micra, the chances would be increasingly good that some key enzymes would find themselves jostled out altogether; with that, growth and multiplication would cease.

Therefore, it is reasonable to suppose that the smallest bacteria visible under a good optical microscope are actually the smallest bits of matter into which all the characteristic processes of life can be squeezed. Such bacteria represent, by this way of thinking, the limit of compactness as far as life is concerned.

But what about organisms still smaller than the smallest bacteria that, lacking some essential enzyme or enzymes, do not, under ordinary conditions, grow and multiply? Granted they are not independently alive, can they yet be considered as fully nonliving?

Before answering, consider that such tiny organisms (which we can call subcells) retain the potentiality of growth and multiplication. The potentiality can be made an actuality once the missing enzyme or enzymes are supplied, and these can only be supplied by a complete and living cell. A subcell, therefore, is an organism that possesses the ability to invade a cell and there, within the cell, to grow and multiply, utilizing the cell's enzymatic equipment to flesh out its own shortcomings.

The largest of the subcells are the rickettsiae, named for an American pathologist, Howard Taylor Ricketts, who, in 1909, discovered that insects were the transmitting agents of Rocky Mountain spotted fever, a disease produced by such subcells. He died the next year of typhus fever, catching it in the course of his researches on that disease, also transmitted by insects. He was thirty-nine at the time of his death; and his reward for giving his life for the good of man is, as you might expect, oblivion.

The smaller rickettsiae fade off into the viruses (there is no sharp dividing line) and the smaller viruses lap over, in size, the genes, which are found in the nuclei of cells and which, in their viruslike structure, carry genetic information.

Now, in considering the subcells, let's abandon the cubic micron as a measure of volume, because if we don't we will run into tiny decimals. Instead, let's use the "cubic millimicron." The millimicron is $\frac{1}{1,000}$ of a micron. A cubic millimicron is, therefore, $\frac{1}{1,000}$ times $\frac{1}{1,000}$ times $\frac{1}{1,000}$, or one-billionth of a cubic micron.

In other words, the smallest bacterium, with a volume of 0.02 cubic micra, can also be said to have a volume of 20,000,000 cubic millimicra. Now we can prepare a table of subcell volumes:

Subcell	Volume (in cubic millimicra)
Typhus fever rickettsia	54,000,000
Cowpox virus	5,600,000
Influenza virus	800,000
Bacteriophage	520,000
Tobacco mosaic virus	50,000
Gene	40,000
Yellow-fever virus	5,600
Hoof-and-mouth virus	700

The range of subcells is huge. The largest rickettsia is nearly three times the size of the smallest bacterium. (It is not size alone that makes an organism a subcell; it is the absence of at least one essential enzyme.) The smallest subcell, on the other hand, is only $\frac{1}{3,500}$ as large as the smallest one as the largest whale is to the average dog.

As one slides down the scale of subcells, the number of molecules decreases. Naturally, the nitrogen-phosphorus macromolecules don't disappear entirely, for life, however distantly potential, is impossible (in the form we know) without them. The very smallest subcells consist of nothing more than a very few of these macromolecules; only the bare essentials of life, so to speak, stripped of all superfluity.

The number of atoms, however, is still sizable. A cubic millimicron will hold several hundred atoms if they were packed with the greatest possible compactness, but of course, in living tissue, they are not.

Thus, the tobacco mosaic virus has a molecular weight of 40,000,000 and the atoms in living tissue have an atomic weight that averages about 8. (All but the hydrogen atom have atomic weights that are well above 8, but the numerous hydrogen atoms, each with an atomic weight of 1, pulls the average far down.)

This means there are roughly 5,000,000 atoms in a tobacco mosaic virus particle, or just about 100 atoms per cubic millimicron. We can therefore prepare a new version of the previous table:

Subcell	Number of atoms
Typhus fever rickettsia	5,400,000,000
Cowpox virus	560,000,000
Influenza virus	80,000,000
Bacteriophage	52,000,000
Tobacco mosaic virus	5,000,000
Gene	4,000,000
Yellow-fever virus	560,000
Hoof-and-mouth virus	70,000

It would seem, then, that the barest essentials of life can be packed into as few as 70,000 atoms. Below that level, we find ordinary protein molecules, definitely non-living. Some protein molecules (definitely nonliving) actually run to more than 70,000 atoms, but the average such molecule contains 5,000 to 10,000 atoms.

Let's consider 70,000 atoms, then, as the "minimum life unit." Since an average human cell contains macromolecules possessing a total number of atoms at least half a billion times as large as the minimum life unit, and since the cerebral cortex of man contains ten billion such cells, it is not at all surprising that our brain is what it is.

In fact, the great and awesome wonder is that mankind, less than ten thousand years after inventing civilization, has managed to put together a mere few thousand excessively simple units and build computers that do as well as they do.

Imagine what would happen if we could make up units containing half a billion working parts, and then use ten billion of those units to design a computer. Why, we would have something that would make the human brain look like a wet firecracker.

Present company excepted, of course!

3 That's Life!

My son is fiendishly interested in outer space. This is entirely without reference to his father's occupation, concerning which he is possessed of complete apathy. Anyway, in honor of this interest of his, we once bought a recording of a humorous skit entitled "The Astronaut" (which was soon worn so thin as the result of repeated playings, that the needle delivered both sides simultaneously).

At one point in this recording, the interviewer asks the astronaut whether he expects to find life on Mars, and the astronaut answers thoughtfully, "Maybe. . . . If I land on Saturday night."

Which brings us face to face with the question of what, exactly, do we mean by life. And we don't have to go to Mars to be faced with a dilemma. There is room for heated arguments right here on earth.

We all know, or think we know, purely on the basis of intuition, what life is. We know that we are alive and that a dead man isn't, that an oyster is alive and a rock isn't. We are also quite confident that such diverse things as sea anemones, gorillas, chestnut trees, sponges, mosses, tape worms and chipmunks are all alive—except when they're dead.

The difficulty arises when we try to take this intuitive knowledge and fit it into words, and this is what I am going to try to do in this chapter.

There is more than one fashion in which we can construct a definition. For instance, we can make a functional definition, or we can make a structural one.

Thus, a child might say: "A house is something to live

in" (functional). Or he might say: "A house is made of brick" (structural).

Neither definition is satisfactory since a tent is something to live in and yet is not ordinarily considered a house, while a wall may be made of brick and yet not be a house.

Combining the two types of definitions may leave it imperfect even so, but it will represent an improvement. Thus "a house is something made of brick in which people live" at once eliminates tents and walls. (It also eliminates frame houses, to say nothing of brick houses that are owned by families who have just left for a month's vacation in the mountains.)

This line of reasoning has an application to definitions involving the concept of life. For instance, when I went to school, the definition I saw most often was functional and went something like this: "A living organism is charterized by the ability to sense its environment and respond appropriately to ingest food, digest it, absorb it, assimilate it, break its substance down and utilize the energy so derived, excrete wastes, grow and reproduce." (When I refer to this later in the chapter, I shall signify the list by "sense its environment, etc.," to save wear and tear on my typewriter ribbon and your retina.)

There was always a question, though, as to whether this was really an exclusive definition. Inanimate objects could imitate these functions if we wanted to argue subtly enough. Crystals grow, for instance, and if we consider the saturated solution to be its food, we certainly might make out a case for absorption and assimilation. Fires can be said to digest their fuel and to leave wastes behind, and they certainly grow and reproduce. Then, too, very simple robots have already been constructed that can imitate all these functions of life (except growth and reproduction) by means of a photocell and wheels.

I tried to define life functionally in another fashion a book entitled *Life and Energy* (Doubleday, 1962). I introduced thermodynamic concepts and said: "A living organism is characterized by the ability to effect a temporary and local decrease in entropy."

As it stands, however, this definition is perfectly terrible, for the sun's heat can also bring about a temporary and local decrease in entropy; for example, every time it evaporates a puddle of water. However, as I shall explain later in the chapter, I don't let this statement stand unmodified. (Incidentally, if you want to know about entropy, I refer you to Chapter 10.)

What we need, clearly, is to introduce something structural into the definition, but can we? All forms of life, however diverse in appearance, have functions in common. They all sense their environment, etc., which is why a functional definition can be set up so easily. But do they all have any structure in common? The mere fact that I use the clause "however diverse in appearance" would indicate they do not.

That, however, is only true if we were to rely on the diversity of ordinary appearance as visible to (if you will excuse the expression) the naked eye. But suppose we clothe the eye in an appropriate lens?

Back in 1665, an English scientist, Robert Hooke, published a book in which he described his researches with a microscope. As part of his research, he studied a thin section of cork and found it to be riddled with tiny rectangular holes. He named the holes "cells," this meaning any small room and therefore being a graphically appropriate word.

But cork is dead tissue even when it occurs on a living tree. Over the next century and a half, microscopists studied living tissue, or, at least, tissue that was alive until they prepared it for study. They found that such tissue was also marked off into tiny compartments The name "cell" was kept for those even though, in living tissue, such compartments were no longer empty holes but were, to all appearances, filled with matter.

It wasn't until the 1830's, though, that accumulating evidence made it possible for two German biologists, Matthias Jakob Schleiden and Theodor Schwann, to present the world with the generalization that all living organisms were made up of cells.

Here, then, is a structural definition: "A living organism is made up of cells."

However, such a definition, although it sounds good, cannot be reversed. You cannot say that an object composed of cells is living, since a dead man is made up of cells just as surely as a living man is, except that the cells of a dead man are dead.

And it does no good to amend the definition by saying that a living organism is composed of living cells, because that is arguing in a circle. Besides, in an organism that is freshly dead, many cells are still alive. Perhaps even the vast majority are—yet the organism is dead.

We can do better, as in the case of the definition of the house, if we include both structural and functional aspects in the definition and say: "A living organism is made up of cells *and* is characterized by the ability to sense its environment, etc."

Here is a definition that includes all the diverse types of organisms we intuitively recognize as living, and excludes everything else, such as crystals, river deltas, flames, robots, and abstractions which can be said to mimic the functions we associate with life—simply because these latter objects do not consist of cells. The definition also excludes dead remnants of once-living objects (however freshly dead) because such dead objects, while constructed of cells, do not perform the functions we associate with life.

I referred in passing, some paragraphs ago, to "living cells." What does that mean?

The definition of a living organism as I have just presented it says that it is made up of cells, but does that imply that the cells themselves are alive? Can we argue that all parts of a living body are necessarily alive and that cells therefore must be alive, as long as they are part of a living organism?

This is clearly a mistaken argument. Hair is not alive, though it is growing on your body. Your skin is covered with a layer of cells that are quite dead by any reasonable criterion though they are part of a living organism.

If we are going to decide whether cells are alive, we can't allow it to depend secondarily on a definition of a living organism. We must apply the necessary criteria of life to the cell itself and ask whether it can sense its environment, etc., and meet, at least, the functional definition of a living thing.

At once the answer is No. Many cells clearly lack one or more of the vital functional abilities of living things. The cells of our nervous system, for instance, cannot reproduce. We are born with the total number of nerve cells we will ever have; any change thereafter can only be for the worse, for a nerve cell that loses its function cannot be replaced.

To be more general, none of our cells, if detached from its neighbors, and set up in business for itself, can long survive to fulfill its functions.

And yet there are different cells among those of our body that can, in themselves, perform each of the functions associated with life. Some cells can sense their environment; others respond appropriately; some surpervise digestion; others absorb; all cells assimilate and produce and use energy; some cells grow and reproduce continually throughout life even after the organism as a whole has ceased to grow and reproduce. In short, the functions of a living organism are, in a sense, the sum of the functions of the cells making it up.

We can say then: "A living cell is one that contributes in some active fashion to the functioning of the organism of which it is a part." This raises the question of what we mean by "some active fashion," but I will leave that to your intuition and say only that the definition is intended to eliminate the problem of the dead cells of the skin, which serve our body only by being there as protection, and not by doing anything actively. Furthermore, a cell many continue its accustomed activities for a limited time after the death of the organisms, and then we can speak of living cells in a dead body.

But there is still an important point to make. We now have two different definitions, one for a living cell and one for a living organism. That means that a cell of a human being is not alive in the same sense that an entire human being is alive. And this makes sense at that, for though the functions of a human being may be viewed as the sum of the functions of his cells, the life of a human being is still more than the sum of the life of his cells.

If you can imagine all the cells of the human body alive in isolation and put together at random, you know that no human being will result. A human being consists not only of something material (cells), but of something rather abstract as well (a specific cell organization). It is quite possible to end human life by destroying the organization while scarcely touching any of the cells themselves.

But I am talking about human cells—is this true for the cells of other organisms as well? Yes, it is; at least, for any reasonably complex organism.

However, as one descends to simpler and simpler organisms, the factor of cellular organization becomes progressively less important. That is, disruption of organization can become more and more extensive without actually putting an end to life. We can replace a lost fingernail, but a lobster can replace a lost limb, A starfish can be cut into sizable chunks and each piece will grow back the remainder, while a sponge can be divided into separate cells which will then reclump and reorganize. At no point, however, is organization of zero importance, as long as an organism consists of more than one cell.

But organisms made up of a single cell do indeed exist, having first been discovered by a Dutch microscopist, Anton van Leeuwenhoek, at the same time that Hooke was discovering cells. A one-celled organism, such as an amoeba, fulfills all the functional requirements of a living organism, in that it can sense its environment, etc. And yet it does not meet the structural portion of the definition, for it is not composed of cells. It *is* a cell.

So we can modify the definition: "A living organism is made up of one or more cells and is characterized by the ability to sense its environment, etc."

It follows, then, that cell organization is not an absolute requirement for *all* types of living organisms. Only the existence of the cell itself seems to be required for the existence of a living thing.

For this reason, it grew popular in the nineteenth century to say that the cell was the "unit of life," and for biologists to devote more and more of their effort toward an understanding of the cell.

But now we can raise the question as to whether the cell actually is the irreducible unit of life or whether something still simpler exists that will serve in that respect.

First, what is a cell? Roughly speaking, we can speak of it as an object that contains at least three parts. First, it possesses a thin membrane that marks it off from the outside universe. Second, it possesses a small internal structure called a nucleus. Third, between membrane and nucleus lies the cytoplasm.

To be sure, there are human cells (such as those of the heart) that run together and are not properly separated by membranes. There are also human cells, such as the red blood corpuscles, that have no nuclei. These are, however, highly specialized cells of a multicellular organism which, in isolation, we cannot consider living organisms.

For those cells that are truly living organisms, it remains true that the membrane, cytoplasm and nucleus are minimum essentials. Some particularly simple one-celled organisms appear to lack nuclei, the bacteria and the blue-green algae being examples. These cells, however, contain "nuclear material"; that is, regions which react chemically as do the intact nuclei of more complicated cells. These simple cells still have nuclei then, but nuclei that are spread through the body of the cell rather than collected in one spot.

Is any one of these parts of the cell more essential than the other two? That may seem like asking which leg of a three-legged stool is more essential, since no cell can live without all three. Nevertheless there is evidence pointing to a gradation of importance. If an amoeba, for instance, is divided by means of a fine needle into two halves, one of which contains the intact nucleus, the half with the nucleus can recover, survive, grow and reproduce normally. The half without the nucleus may carry on the functions of life for a short while, but cannot grow or reproduce.

Furthermore, when a cell divides, it goes through a complicated series of changes that particularly involve small structures called chromosomes, which lie within the nucleus. This is true whether a cell is an organism in itself or is merely part of a larger organism.

The changes in which the chromosomes are involved include a key step, one in which each chromosome induces the formation of another like itself. This is called "replication," for the chromosome has produced a replica of itself. No cell ever divides without replication taking place. As the nineteenth century drew to an end, the suspicion began to stir in biologists that as the cell was the key to the organism, so the chromosome was the key to the cell.

We can help matters along if we turn once again to the structural definition. After all, our definition of a living organism is both functional and structural as far as multicellular organisms are concerned. They are composed of cells. For a one-celled organism, the definition becomes purely functional, for there is nothing to say what a single cell is composed of.

To clarify that point, we can descend to the molecular level. A cell contains numerous types of molecules, some of which are also to be found in inanimate nature and which are, therefore, however characteristic of living organisms, not characteristic *only* of living organisms. (Water is an example.)

Yet there are molecules that are to be found only in living cells or in material that was once part of a living cell or, at the very least, had been formed by a living cell. The most characteristic of these (for reasons I won't go into here) are the various protein molecules. No form of life exists, no single cell, however simple or however complicated, that does not contain protein.

Proteins satisfy a variety of functions. Some merely make up part of the substratum of the body, forming major components of skin, hair, cartilage, tendons, ligaments and so on. Other proteins, however, are intimately concerned with the actual chemical workings of the cell; they catalyze the thousands of reactions that go on. These proteins (called enzymes) are, we cannot help but intuitively feel, close to the chemical essence of life.

In fact, I can now return to my book *Life and Energy*, from which I quoted an unsatisfactory definition of the living organism near the beginning of the chapter, and can explain how I amended the definition to make it satisfactory, thus: "A living organism is characterized by the ability to effect a temporary and local decrease in entropy by means of enzyme-catalyzed reactions." Here is a definition that is both functional (it effects an entropy decrease) and structural (by means of enzymes).

Now, this definition *does not involve cells*. It applies as truly to a multicellular as to a unicellular organism, and it accurately marks off those systems we intuitively recognize as alive from those we do not.

This new definition would make it seem that it is not the cell so much that is the unit of life, but the enzymes within the cell. However, if enzymes can only be formed within cells and by cells, the distinction is a purely academic one. Unless, that is, we can pin down the manufacture of enzymes to something more specific than the cell as a whole.

In recent decades it has become quite obvious that the thousands of different enzymes present in each cell (one for each of the thousands of different chemical reactions that are continually proceeding within the cell) are formed under the supervision of the chromosomes.

Shifting to the chromosomes then, and remaining on the molecular level, I must explain that the chromosomes are composed of a series of giant protein molecules of a variety called nucleoprotein, because each consists of a protein portion and a nucleic acid portion. The nucleic acid portion is quite different from the protein in structure.

Nucleic acid is so named because it was found originally, in the nucleus. Since the first days, it has also been found in the cytoplasm, but it keeps its original name. There are two forms of nucleic acid, with complicated names that are abbreviated DNA and RNA. DNA is found in the nucleus and makes up a major portion of the chromosomes. RNA is found chiefly in the cytoplasm, though a small quantity is also present in the nucleus.

Research in the 1950's has shown that it is not merely the chromosomes but the DNA content thereof (with an assist from RNA) that supervises the synthesis of specific enzymes. Through those enzymes, the nucleic acids of the cell might be said to supervise the chemical activity of the cell and to be, therefore, in control of all the functions we associate with living organisms.

But though nucleic acids control the functions of living organisms, can they themselves be considered "living"? When this question arose earlier in the chapter in connection with cells, I wasn't satisfied that a cell was truly alive until it could be shown that a single cell could serve as an organism in itself. Similarly, we can't consider nucleic acids to be alive until and unless we can show that a nucleic acid molecule can serve as an organism in itself.

Let's go back in time again.

Back in the 1880's the French biochemist Louis Pasteur, while studying hydrophobia, tried to isolate the germ of the disease. Twenty years earlier, you see, he had evolved the "germ theory of disease," which stated that all infectious diseases were caused and transmitted by microorganisms. Hydrophobia was certainly infectious, but where was the microorganism?

Pasteur had two choices. He could abandon his theory or he could introduce an *ad hoc* amendment (that is, one designed for no other purpose than to explain away a specific difficulty). Ordinarily the introduction of *ad hoc* amendments is a poor procedure, but a genius can get away with it. Pasteur suggested that the germ of hydrophobia existed, but was too small to be seen in a microscope.

Pasteur was right.

Another disease studied at the time, by botanists, was tobacco mosaic disease, one in which the leaves of tobacco plants were mottled into a mosaic. The juice from a diseased leaf would infect a healthy leaf, so by Pasteur's theory, a germ should exist. None, however, could be found here, either.

In 1892, a Russian bacteriologist, Dmitri Ivanovski, ran some of the juice of a diseased leaf through a porcelain filter that was so fine that no bacterium, not even the smallest, could pass through. The juice that did get through was still capable of passing on the disease. The infectious agent was therefore called a "filtrable virus." ("Virus" simply means "poison," so a "filtrable virus" is a poison that passes through a filter.)

Other diseases, including hydrophobia, were found to be transmitted by filtrable viruses. The nature of these viruses, however, was unknown until 1931, when an English bacteriologist, William J. Elford, designed a filter fine enough to trap the virus. In this way the virus, though smaller by far than even the smallest cells, proved to be larger by far than most molecules.

Well, then, was the virus particle (whatever its nature) a living organism? It infects cells so it must somehow sense their presence and respond appropriately. It must feed on their substance, absorb, assimilate, make use of energy, grow and reproduce. And yet the virus particle certainly did not consist of cells as they were then known. The whole problem of the nature of life was thrown into confusion in the 1930's, although the problem had been clarified by the cell theory in the 1830's.

In 1935, the American biochemist Wendell Meredith Stanley actually succeeded in crystallizing the tobacco mosaic virus, and this seemed to be a forceful argument against life. Even after the virus had been crystallized it remained infective, and how can anything living survive crystallization, for goodness' sake? Crystals were objects associated only with nonliving things.

Actually, this argument is worthless. Nothing alive can be crystallized because, until viruses were discovered, nothing alive was simple enough to be crystallized. But viruses were simpler than any cellular form of life and there was no reason in the world to suppose that the non-crystallization rule ought to apply to them.

Once enough tobacco mosaic virus was purified and brought together by crystallization, it could be tested chemically, and it was found by two British biochemists, Frederick C. Bawden and Norman W. Pirie, to be a nucleoprotein. It was 94 percent protein and 6 percent RNA.

Since then, without exception, all viruses that have been analyzed have proved to be nucleoprotein. Some contain DNA, some RNA, some both—but none are completely lacking in nucleic acid.

Furthermore, when a virus infects a cell, it is the nucleic acid portion that actually enters the cell, while the protein portion remains outside. There is now every reason to think that the protein is merely a nonliving shell about the nucleic acid, which is itself the key portion of the virus. Naked nucleic acid molecules have even been prepared from viruses and, in themselves, have remained slightly infective.

It certainly looks as though in the virus we have found our example of a nucleic acid molecule that in itself and by itself behaves as a living organism.

Suppose we say, then: "A living organism is characterized by the possession of at least one molecule of nucleic acid capable of replication." This definition is both structural (the nucleic acid) and functional (the replication). It includes not only all cellular life, but all viruses as well; and it excludes all things else.

To be sure, there are arguments against this. Some feel that the virus is not a true example of a living organism because it cannot perform its function until it is inside the cell. Within the cell, and only within the cell, does it supervise enzyme action and bring about the synthesis of specific enzymes and other proteins. It does this by making use of the cell's chemical machinery, including its enzymes. Outside the cell, the virus performs none of the functions we associate with life. The cell, therefore, so this argument goes, is still the unit of life.

I do not see the force of this argument. To be sure, the virus requires a cell in order to perform certain of its functions, but its life outside the cell is not wholly static. It must actively penetrate the cell, and must do that without the help of the cell itself. This is an example of at least one action characteristic of life (the equivalent of ingestion of food, somewhat inside-out) that it performs all by itself.

Then, even if we admit that the virus makes use of cellular machinery for some of its functions, so does a tapeworm make use of our cellular machinery for some of its functions. The virus, like the tapeworm, is a parasite, but happens to be a more complete one. Shall we draw an artificial line and say that the tapeworm is a living organism and the virus is not?

Furthermore, all organisms, parasites or not, are as dependent upon some factor of the outer world as viruses are. We ourselves, for instance, could not live for more than a few minutes if our access to oxygen were cut off. Is that any reason to suppose that we are not living organisms but that it is the oxygen that is really the living organism? Why, therefore, put the necessary outside cell (for the virus) in a category different from that of the necessary outside oxygen (for us)?

Nor is there anything crucial in the fact that the virus makes use of enzymes that are not its own. Let me explain this point by analogy.

Consider the woodcutter chopping down a tree with an ax. He can't do it without an ax, and yet we never think of a woodcutter as a man-ax combination. A woodcutter is a man and the ax is merely the woodcutter's tool. Similarly, a nucleic acid may not be able to perform its actions without enzymes, but the enzymes are merely its tool, while the nucleic acid is the thing itself.

Furthermore, when a woodcutter is in action, chopping down a tree, the ax may be his or it may be stolen. This may make him an honest man or a thief, respectively, but in either case, he is a woodcutter in action. In the same way, a virus performing its functions is a living organism whether the enzymes it uses are its own or not.

As far as I am concerned, therefore, my definition of living organisms in terms of their nucleic acid content is a valid one.

It's necessary to remember, of course, that a living organism is more than its nucleic acid content, just as it is more than its cellular content. As I said earlier in the chapter, a living organism consists not only of separate parts, but of those parts in appropriate organization.

There are some biologists who deplore the intense concentration on DNA in contemporary biological and biochemical research. They feel that organization is being neglected in favor of a study of the parts alone, and I must admit there is some justification to this.

Nevertheless, I also feel that we will never understand the organization until we have a thorough understanding of the parts being organized, and it is my hope that when the DNA molecule is laid out plain for all to see, many of the current mysteries of life will fall neatly into place—organization and all.

4 Not as We Know It

Even unpleasant experiences can be inspiring.

For instance, my children once conned me into taking them to a monster-movie they had seen advertised on TV.

"It's *science fiction*," they explained. They don't exactly know what science fiction is, but they have gathered it's something daddy writes, so the argument is considered very powerful.

I tried to explain that it wasn't science fiction by *my* definition, but although I had logic on my side, they had decibels on theirs.

So I joined a two-block line consisting of every kid for miles around with an occasional grown-up who spent his time miserably pretending he was waiting for a bus and would leave momentarily. It was a typical early spring day in New England—nasty drizzle whipped into needle-spray by a howling east wind—and we inched slowly forward.

Finally, when we were within six feet of the ticket-sellers and I, personally, within six inches of pneumonia, my guardian angel smiled and I had my narrow escape. They hung up the SOLD OUT sign.

I said, with a merry laugh, "Oh, what a dirty shame," and drove my howlingly indignant children home.

Anyway, it got me to thinking about the lack of imagination in movieland's monsters. Their only attributes are their bigness and destructiveness. They include big apes, big octopuses (or is the word "octopodes"?), big eagles, big spiders, big amoebae. In a way, that is all Hollywood needs, I suppose. This alone suffices to drag in huge crowds of vociferous human larvae, for to be big and destructive

49

is the secret dream of every red-blooded little boy and girl in the world.

What, however, is mere size to the true *aficionado?* What we want is real variety. When the cautious astronomer speaks of life on other worlds with the qualification "life-as-we-know-it," we become impatient.

What about life-not-as-we-know-it?

Well, that's what I want to discuss in this chapter.

To begin with, we have to decide what life-as-we-know-it means. Certainly life-as-we-know-it is infinitely various. It flies, runs, leaps, crawls, walks, hops, swims, and just sits. It is green, red, yellow, pink, dead white and vari-colored. It glows and does not glow, eats and does not eat. It is boned, shelled, plated and soft; has limbs, tentacles or no appendages at all; is hairy, scaly, feathery, leafy, spiny and bare.

If we're going to lump it all as life-as-we-know-it, we'll have to find out something it all has in common. We might say it is all composed of cells, except that this is not so. The virus, an important life form to anyone who has ever had a cold, is not.

So we must strike beyond physiology and reach into chemistry, saying that all life is made up of a directing set of nucleic acid molecules which controls chemical reactions through the agency of proteins working in a watery medium.

There is more, almost infinitely more, to the details of life, but I am trying to strip it to a basic minimum. For life-as-we-know-it, water is the indispensable background against which the drama is played out, and nucleic acids and proteins are the featured players.

Hence any scientist, in evaluating the life possibilities on any particular world, instantly dismisses said world if it lacks water; or if it possesses water outside the liquid range, in the form of ice only or of steam only.

(You might wonder, by the way, why I don't include oxygen as a basic essential. I don't because it isn't. To be sure, it is the substance most characteristically involved in the mechanisms by which most life forms evolve energy, but it is not invariably involved. There are tissues in our body that can live temporarily in the absence of molecular oxygen, and there are microorganisms that can live indefinitely in the absence of oxygen. Life on earth almost certainly developed in an oxygen-free atmosphere, and even today there are microorganisms that can live *only* in the absence of oxygen. No known life form on earth, however, can live in the complete absence of water, or fails to contain both protein and nucleic acid.)

In order to discuss life-not-as-we-know-it, let's change either the background or the feature players. Background first!

Water is an amazing substance with a whole set of unusual properties which are ideal for life-as-we-know-it. So well fitted for life is it, in fact, that some people have seen in the nature of water a sure sign of Divine providence. This, however, is a false argument, since life has evolved to fit the watery medium in which it developed. Life fits water, rather than the reverse.

Can we imagine life evolving to fit some other liquid, then, one perhaps not too different from water? The obvious candidate is ammonia.

Ammonia is very like water in almost all ways. Whereas the water molecule is made up of an oxygen atom and two hydrogen atoms (H_2O) for an atomic weight of 18, the ammonia molecule is made up of a nitrogen atom and three hydrogen atoms (NH_3) for an atomic weight of 17. Liquid ammonia has almost as high a heat capacity as liquid water, almost as high a heat of evaporation, almost as high a versatility as a solvent, almost as high a tendency to liberate a hydrogen ion.

In fact, chemists have studied reactions proceeding in liquid ammonia and have found them to be quite analogous to those proceeding in water, so that an "ammonia chemistry" has been worked out in considerable detail.

Ammonia as a background to life is therefore quite conceivable—but not on earth. The temperatures on earth are such that ammonia exists as a gas. Its boiling point at atmospheric pressure is −33.4° C. (−28° F.) and its freezing point is −77.7° C. (−108° F.).

But other planets?

In 1931, the spectroscope revealed that the atmosphere

of Jupiter, and, to a lesser extent, of Saturn, was loaded with ammonia. The notion arose at once of Jupiter being covered by huge ammonia oceans.

To be sure, Jupiter may have a temperature not higher than −100° C. (−148° F.), so that you might suppose the mass of ammonia upon it to exist as a solid, with atmospheric vapor in equilibrium. Too bad. If Jupiter were closer to the sun . . .

But wait! The boiling point I have given for ammonia is at atmospheric pressure—earth's atmosphere. At higher pressures, the boiling point would rise, and if Jupiter's atmosphere is dense enough and deep enough, ammonia oceans might be possible after all. (Other points of view are also possible—see Chapter 16.)

An objection that might, however, be raised against the whole concept of an ammonia background for life, rests on the fact that living organisms are made up of unstable compounds that react quickly, subtly and variously. The proteins that are so characteristic of life-as-we-know-it must consequently be on the edge of instability. A slight rise in temperature and they break down.

A drop in temperature, on the other hand, might make protein molecules too stable. At temperatures near the freezing point of water, many forms of non-warm-blooded life become sluggish indeed. In an ammonia environment, with temperatures that are a hundred or so Centigrade degrees lower than the freezing point of water, would not chemical reactions become too slow to support life?

The answer is twofold. In the first place, why is "slow" to be considered "too slow"? Why might there not be forms of life that live at slow motion compared to ourselves? Plants do.

A second and less trivial answer is that the protein structure of developing life adapted itself to the temperature by which it was surrounded. Had it adapted itself over the space of a billion years to liquid ammonia temperatures, protein structures might have been evolved that would be far too unstable to exist for more than a few minutes at liquid water temperatures, but are just stable enough to exist conveniently at liquid ammonia temperatures. These new forms would be just stable enough and unstable enough at low temperatures to support fast-moving forms of life.

Nor need we be concerned over the fact that we can't imagine what those structures might be. Suppose we were creatures who lived constantly at a temperature of a dull red heat (naturally with a chemistry fundamentally different from that we now have). Could we under those circumstances know anything about earth-type proteins? Could we refrigerate vessels to a mere 25° C., form proteins and study them? Would we ever dream of doing so, unless we first discovered life forms utilizing them?

Anything else besides ammonia now?

Well, the truly common elements of the universe are hydrogen, helium, carbon, nitrogen, oxygen and neon. We eliminate helium and neon because they are completely inert and take part in no reactions. In the presence of a vast preponderance of hydrogen throughout the universe, carbon, nitrogen and oxygen would exist as hydrogenated compounds. In the case of oxygen, that would be water (H_2O), and in the case of nitrogen, that would be ammonia (NH_3). Both of these have been considered. That leaves carbon, which, when hydrogenated, forms methane (CH_4).

There is methane in the atmosphere of Jupiter and Saturn, along with ammonia; and, in the still more distant planets of Uranus and Neptune, methane is predominant, as ammonia is frozen out. This is because methane is liquid over a temperature range still lower than that of ammonia. It boils at −161.6° C. (−259° F.) and freezes at −182.6° C. (−297° F.) at atmospheric pressure.

Could we then consider methane as a possible background to life with the feature players being still more unstable forms of protein? Unfortunately, it's not that simple.

Ammonia and water are both polar compounds; that is, the electric charges in their molecules are unsymmetrically distributed. The electric charges in the methane molecule are symmetrically distributed, on the other hand, so it is a non-polar compound.

Now, it so happens that a polar liquid will tend to dissolve polar substances but not nonpolar substances, while a nonpolar liquid will tend to dissolve nonpolar substances but not polar ones.

Thus water, which is polar, will dissolve salt and sugar, which are also polar, but will not dissolve fats or oils (lumped together as "lipids" by chemists), which are nonpolar. Hence the proverbial expression, "Oil and water do not mix."

On the other hand, methane, a nonpolar compound, will dissolve lipids but will not dissolve salt or sugar.

Proteins and nucleic acids are polar compounds and will not dissolve in methane. In fact, it is difficult to conceive of any structure that would jibe with our notions of what a protein or nucleic acid ought to be that would dissolve in methane.

If we are to consider methane, then, as a background for life, we must change the feature players.

To do so, let's take a look at protein and nucleic acid and ask ourselves what it is about them that makes them essential for life.

Well, for one thing, they are giant molecules, capable of almost infinite variety in structure and therefore potentially possessed of the versatility required as the basis of an almost infinitely varying life.

Is there no other form of molecule that can be as large and complex as proteins and nucleic acids and that can be nonpolar, hence soluble in methane, as well? The most common nonpolar compounds associated with life are the lipids, so we might ask if it is possible for there to exist lipids of giant molecular size.

Such giant lipid molecules are not only possible; they actually exist. Brain tissue, in particular, contains giant lipid molecules of complex structure (and of unknown function). There are large "lipoproteins" and "proteolipids" here and there which are made up of both lipid portions and protein portions combined in a single large molecule. Man is but scratching the surface of lipid chemistry; the potentialities of the nonpolar molecule are greater than we have, until recent decades, realized.

Remember, too, that the biochemical evolution of earth's life has centered about the polar medium of water.

Had life developed in a nonpolar medium, such as that of methane, the same evolutionary forces might have endlessly proliferated lipid molecules into complex and delicately unstable forms that might then perform the functions we ordinarily associate with proteins and nucleic acids.

Working still further down on the temperature scale, we encounter the only common substances with a liquid range at temperatures below that of liquid methane. These are hydrogen, helium, and neon. Again, eliminating helium and neon, we are left with hydrogen, the most common substance of all. (Some astronomers think that Jupiter may be four-fifths hydrogen, with the rest mostly helium—in which case good-by ammonia oceans after all.)

Hydrogen is liquid between temperatures of −253° C. (−423° F.) and −259° C. (−434° F.), and no amount of pressure will raise its boiling point higher than −240° C. (−400° F.). This range is only twenty to thirty Centigrade degrees over absolute zero, so that hydrogen forms a conceivable background for the coldest level of life. Hydrogen is nonpolar, and again it would be some sort of lipid that would represent the featured player.

So far the entire discussion has turned on planets colder than the earth. What about planets warmer?

To begin with, we must recognize that there is a sharp chemical division among planets. Three types exist in the solar system and presumably in the universe as a whole.

On cold planets, molecular movements are slow, and even hydrogen and helium (the lightest and therefore the nimblest of all substances) are slow-moving enough to be retained by a planet in the process of formation. Since hydrogen and helium together make up almost all of matter, this means that a large planet would be formed. Jupiter, Saturn, Uranus and Neptune are the examples familiar to us.

On warmer planets, hydrogen and helium move quickly enough to escape. The more complex atoms, mere impurities in the overriding ocean of hydrogen and helium, are sufficient to form only small planets. The chief hydrogenated compound left behind is water, which is the high-

55

est-boiling compound of the methane-ammonia-water trio and which, besides, is most apt to form tight complexes with the silicates making up the solid crust of the planet.

Worlds like Mars, earth, and Venus result. Here, ammonia and methane forms of life are impossible. Firstly, the temperatures are high enough to keep those compounds gaseous. Secondly, even if such planets went through a super-ice-age, long aeons after formation, in which temperatures dropped low enough to liquefy ammonia or methane, that would not help. There would be no ammonia or methane in quantities sufficient to support a world-girdling life form.

Imagine, next, a world still warmer than our medium trio: a world hot enough to lose even water. The familiar example is Mercury. It is a solid body of rock with little, if anything, in the way of hydrogen or hydrogen-containing compounds.

Does this eliminate any conceivable form of life that we can pin down to existing chemical mechanisms?

Not necessarily.

There are nonhydrogenous liquids, with ranges of temperature higher than that of water. The most common of these, on a cosmic scale, would be sulfur which, under one-atmosphere pressure, has a liquid range from 113° C. (235° F.) to 445° C. (833 ° F.); this would fit nicely into the temperature of Mercury's sunside.

But what kind of featured players could be expected against such a background?

So far all the complex molecular structures we have considered have been ordinary organic molecules; giant molecules, that is, made up chiefly of carbon and hydrogen, with oxygen and nitrogen as major "impurities" and sulfur and phosphorus as minor ones. The carbon and hydrogen alone would make up a nonpolar molecule; the oxygen and nitrogen add the polar qualities.

In a watery background (oxygen-hydrogen) one would expect the oxygen atoms of tissue components to outnumber the nitrogen atoms, and on earth this is actually so. Against an ammonia background, I imagine nitrogen atoms would heavily outnumber oxygen atoms. The two subspecies of proteins and nucleic acids that result might be differentiated by an O or an N in parentheses, indicating which species of atom was the more numerous.

The lipids, featured against the methane and hydrogen backgrounds, are poor in both oxygen and nitrogen and are almost entirely carbon and hydrogen, which is why they are nonpolar.

But in a hot world like Mercury, none of these types of compounds could exist. No organic compound of the types most familiar to us, except for the very simplest, could long survive liquid sulfur temperatures. In fact, earthly proteins could not survive a temperature of 60° C. for more than a few minutes.

How then to stabilize organic compounds? The first thought might be to substitute some other element for hydrogen, since hydrogen would, in any case, be in extremely short supply on hot worlds.

So let's consider hydrogen. The hydrogen atom is the smallest of all atoms and it can be squeezed into a molecular structure in places where other atoms will not fit. Any carbon chain, however intricate, can be plastered round and about with small hydrogen atoms to form "hydrocarbons." Any other atom, but one, would be too large.

And which is the "but one"? Well, an atom with chemical properties resembling those of hydrogen (at least as far as the capacity for taking part in particular molecular combinations is concerned) and one which is almost as small as the hydrogen atom, is that of fluorine. Unfortunately, fluorine is so active that chemists have always found it hard to deal with and have naturally turned to the investigation of tamer atomic species.

This changed during World War II. It was then necessary to work with uranium hexafluoride, for that was the only method of getting uranium into a compound that could be made gaseous without trouble. Uranium research had to continue (you know why), so fluorine had to be worked with, willy-nilly.

As a result, a whole group of "fluorocarbons," complex molecules made up of carbon and fluorine rather than carbon and hydrogen, were developed, and the basis laid for a kind of fluoro-organic chemistry.

To be sure, fluorocarbons are far more inert than the

corresponding hydrocarbons (in fact, their peculiar value to industry lies in their inertness) and they do not seem to be in the least adaptable to the flexibility and versatility required by life forms.

However, the fluorocarbons so far developed are analogous to polyethylene or polystyrene among the hydro-organics. If we were to judge the potentialities of hydro-organics only from polyethylene, I doubt that we would easily conceive of proteins.

No one has yet, as far as I know, dealt with the problem of fluoroproteins or has even thought of dealing with it—but why not consider it? We can be quite certain that they would not be as active as ordinary proteins at ordinary temperatures. But on a Mercury-type planet, they would be at higher temperatures, and where hydro-organics would be destroyed altogether, fluoro-organics might well become just active enough to support life, particularly the fluoro-organics that life forms are likely to develop.

Such fluoro-organic-in-sulfur life depends, of course, on the assumption that on hot planets, fluorine, carbon and sulfur would be present in enough quantities to make reasonably probable the development of life forms by random reaction over the life of a solar system. Each of these elements is moderately common in the universe, so the assumption is not an altogether bad one. But, just to be on the safe side, let's consider possible alternatives.

Suppose we abandon carbon as the major component of the giant molecules of life. Are there any other elements which have the almost unique property of carbon—that of being able to form long atomic chains and rings—so that giant molecules reflecting life's versatility can exist?

The atoms that come nearest to carbon in this respect are boron and silicon, boron lying just to the left of carbon on the periodic table (as usually presented) and silicon just beneath it. Of the two, however, boron is a rather rare element. Its participation in random reactions to produce life would be at so slow a rate, because of its low concentration in the planetary crust, that a boron-based life formed within a mere five billion years is of vanishingly small probability.

That leaves us with silicon, and there, at least, we are on firm ground. Mercury, or any hot planet, may be short on carbon, hydrogen and fluorine, but it must be loaded with silicon and oxygen, for these are the major components of rocks. A hot planet which begins by lacking hydrogen and other light atoms and ends by lacking silicon and oxygen as well, just couldn't exist because there would be nothing left in enough quantity to make up more than a scattering of nickel-iron meteorites.

Silicon can form compounds analogous to the carbon chains. Hydrogen atoms tied to a silicon chain, rather than to a carbon chain, form the "silanes." Unfortunately, the silanes are less stable than the corresponding hydrocarbons and are even less likely to exist at high temperatures in the complex arrangements required of molecules making up living tissue.

Yet it remains a fact that silicon does indeed form complex chains in rocks and that those chains can easily withstand temperatures up to white heat. Here, however, we are not dealing with chains composed of silicon atoms only (Si-Si-Si-Si-Si) but of chains of silicon atoms alternating with oxygen atoms (Si-O-Si-O-Si).

It so happens that each silicon atom can latch on to four oxygen atoms, so you must imagine oxygen atoms attached to each silicon atom above and below, with these oxygen atoms being attached to other silicon atoms also, and so on. The result is a three-dimensional network, and an extremely stable one.

But once you begin with a silicon-oxygen chain, what if the silicon atom's capacity for hooking on to two additional atoms is filled not by more oxygen atoms but by carbon atoms, with, of course, hydrogen atoms attached? Such hybrid molecules, both silicon- and carbon-based, are the "silicones." These, too, have been developed chiefly during World War II and since, and are remarkable for their great stability and inertness.

Again, given greater complexity and high temperature, silicones might exhibit the activity and versatility necessary for life. Another possibility: Perhaps silicones may exist in which the carbon groups have fluorine atoms attached, rather than hydrogen atoms. Fluorosilicones would

be the logical name for these, though, as far as I know—
and I stand very ready to be corrected—none such have
yet been studied.

Might there possibly be silicone or fluorosilicone life
forms in which simple forms of this class of compound
(which can remain liquid up to high temperatures) might
be the background of life and complex forms the princi-
pal character?

There, then, is my list of life chemistries, spanning the
temperature range from near red heat down to near absolute zero:

1. fluorosilicone in fluorosilicone
2. fluorocarbon in sulfur
3.* nucleic acid/protein (O) in water
4. nucleic acid/protein (N) in ammonia
5. lipid in methane
6. lipid in hydrogen

Of this half dozen, the third only is life-as-we-know-it.
Lest you miss it, I've marked it with an asterisk.

This, of course, does not exhaust the imagination, for
science-fiction writers have postulated metal beings living
on nuclear energy, vaporous beings living in gases, energy
beings living in stars, mental beings living in space, indescribable beings living in hyperspace, and so on.

It does, however, seem to include the most likely forms
that life can take as a purely chemical phenomenon based
on the common atoms of the universe.

Thus, when we go out into space there may be more to
meet us than we expect. I would look forward not only to
our extra-terrestrial brothers who share life-as-we-know-it.
I would hope also for an occasional cousin among the life-
not-as-we-know-it possibilities.

In fact, I think we ought to prefer our cousins. Competition may be keen, even overkeen, with our brothers, for we
may well grasp at one another's planets; but there need
only be friendship with our hot-world and cold-world
cousins, for we dovetail neatly. Each stellar system might
pleasantly support all the varieties, each on its own planet,

and each planet useless to and undesired by any other variety.

How easy it would be to observe the Tenth Commandment then!

II CHEMISTRY

5 The Element of Perfection

In the old days of science fiction, when writers had much more of the leeway that arises out of scientific innocence, a "new element" could always be counted on to get a story going or save it from disaster. A new element could block off gravity, or magnify atoms to visible size, or transport matter.

Much of this "new element" fetish was the outcome of the Curies' dramatic discovery in uranium ore of that unusual element, radium, in 1898. And yet, that same decade, another element was found in uranium ore under even more dramatic circumstances. Though this second element aroused nothing like the furore created by radium, it proved, in the end, to be the most unusual element of all and to have properties as wild as any a science-fictioneer ever dreamed up.

Furthermore, the significance of this element to man has expanded remarkably in the past five years, and thoughts connected with that expansion have brought a remarkable vision to my mind which I will describe eventually.

In 1868, there was a total eclipse of the sun visible in India, and astronomers assembled jubilantly in order to bring to bear a new instrument in their quest for knowledge.

This was the spectroscope, developed in the late 1850's by the German scientists Gustav Robert Kirchhoff and Robert Wilhelm Bunsen. Essentially this involved the conduction of the light emitted by heated elements through a prism to produce a spectrum in which the wave lengths of the light could be measured. Each element produced light of wave lengths characteristic of itself, so that the elements were "fingerprinted."

The worth of this new analytical method was spectacularly demonstrated in 1860 when Kirchhoff and Bunsen heated certain ores, came across spectral lines that did not jibe with those already known and, as a result, discovered a rare element, cesium. The next year, they showed this was no accident by discovering another element, rubidium.

With that record of accomplishment, astronomers were eager to turn the instrument on the solar atmosphere (unmasked only during eclipses) in order to determine its chemical composition across the gulfs of space.

Almost at once, the French astronomer Pierre J. C. Janssen observed a yellow line that did not quite match any known line. The English astronomer Norman Lockyer, particularly interested in spectroscopy, decided this represented a new element. He named it for the Greek god of the sun, Helios, so that the new element became "helium."

So far, so good, except that very few, if any, of the earthly chemists cared to believe in a non-earthly element on the basis of a simple line of light. Lockyer's suggestion was greeted with reactions that ran the gamut from indifference to mockery.

Of course, such conservatism appears shameful in hindsight. Actually, though, hindsight also proves the skepticism to have been justified. A new spectral line does not necessarily signify a new element.

Spurred on by the eventual success of helium, other "new elements" were found in outer space. Strange lines in the spectra of certain nebulae were attributed to an element called "nebulium." Unknown lines in the sun's corona were attributed to "coronium" and similar lines in the auroral glow to "geocoronium."

These new elements, however, proved to be delusions. They were produced by old, well-known elements under strange conditions duplicated in the laboratory only years afterward. "Nebulium" and "geocoronium" turned out to be merely oxygen-nitrogen mixtures under highly ionized conditions. "Coronium" lines were produced by highly ionized metals such as calcium.

So you see that the mere existence of the "helium" line did not really prove the existence of a new element. However, to carry the story on, it is necessary to backtrack still another century to a man even further ahead of his times than Lockyer was.

In 1785, the English physicist Henry Cavendish was studying air, which at the time had just been discovered to consist of two gases, oxygen and nitrogen. Nitrogen was an inert gas; that is, it would not combine readily with other substances, as oxygen would. In fact, nitrogen was remarkable for the number of negative properties it had. It was colorless, odorless, tasteless, insoluble and incombustible. It was not poisonous in itself, but neither would it support life.

Cavendish found that by using electric sparks, he could persuade the nitrogen to combine with oxygen. He could then absorb the resulting compound, nitrogen oxide, in appropriate chemicals. By adding more oxygen he could consume more and more of the nitrogen until finally his entire supply was reduced to a tiny bubble which was about 1 per cent of the original volume of air. This last bubble he could do nothing with, and he stated that in his opinion there was a small quantity of an unknown gas in the atmosphere which was even more inert than nitrogen.

Here was a clear-cut experiment by a first-rank scientist who reached a logical conclusion that represented, we now realize, the pure truth. Nevertheless, Cavendish's work was ignored for a century.

Then, in 1882, the British physicist John William Strutt (more commonly known as Lord Rayleigh, because he happened to be a baron as well as a scientist) was investigating the densities of hydrogen and oxygen gas in order better to determine their atomic weights, and he threw in nitrogen for good measure. To do the job with the proper thoroughness, he prepared each element by several different methods. In the case of hydrogen and oxygen, he got the same densities regardless of the method of preparation. Not so in the case of nitrogen.

He prepared nitrogen from ammonia and obtained a density of 1.251 grams per liter. He also prepared nitrogen from air by removing the oxygen, carbon dioxide and water vapor, and for that nitrogen he obtained a density of 1.257 grams per liter. This discrepancy survived his most careful efforts. Helplessly, he published these results in a scientific journal and invited suggestions from the readers, but none were received. Lord Rayleigh, himself, thought of several possible explanations: that the atmospheric nitrogen was contaminated with the heavier oxygen, or with the triatomic molecule N_3, a kind of nitrogen analog of ozone; or that the nitrogen from the ammonia was contaminated with the lighter hydrogen or with atomic nitrogen. He checked out each possibility and all failed.

Then, a decade later, a Scottish chemist, William Ramsay, came to work for Lord Rayleigh and, tackling the nitrogen problem, harked back to Cavendish and wondered if the atmosphere might not contain small quantities of a gas that remained with nitrogen when everything else was removed and which, being heavier than nitrogen, gave atmospheric nitrogen a spuriously high density.

In 1894, Ramsay repeated Cavendish's original experiment with improvements. He passed atmospheric nitrogen over redhot metallic magnesium. Nitrogen wasn't so inert that it could resist that. It reacted with the metal to form magnesium nitride. But not all of it did. As in Cavendish's case, a small bubble was left which was so inert that even hot magnesium left it cold. Ramsay measured its density and it was distinctly heavier than nitrogen. Well, was it a new element or was it merely N_3, a new and heavy form of nitrogen?

But now the spectroscope existed. The unknown gas was heated and its spectrum was observed and found to have lines that were completely new. The decision was reached at once that here was a new element. It was named "argon" from a Greek word meaning "lazy" because of its refusal to enter into any chemical combinations.

Eventually, an explanation for argon's extraordinary inertness was worked out. Each element is made up of atoms containing a characteristic number of electrons arranged in a series of shells something like the layers of an onion. To picture the situation as simply as possible, an atom is most stable when the outermost shell contains eight electrons. Chemical reactions take place in such a way that an atom either gets rid of a few electrons or takes up a few, achieving, in this way, the desired number of eight.

But what if an element has eight electrons in its outermost shell to begin with? Why, then, it is "happy" and need not react at all—and doesn't. Argon is an example. It has three shells of electrons, with the third and outermost containing eight electrons.

After argon was discovered, other examples of inert gases were located; nowadays, six are known altogether: neon, with two shells of electrons; krypton, with four shells; xenon, with five shells; and radon with six shells. In each case, the outermost shell contains eight electrons. (Krypton, xenon and radon have been found, in 1962, to undergo some chemical reactions, but that is beside the present point.)

But I have mentioned only five inert gases. What of the sixth? Ah, the sixth is helium, so let's take up the helium story again.

Just before the discovery of the inert gases, in 1890, to be exact, the American chemist William Francis Hillebrand analyzed a mineral containing uranium and noticed that it gave off small quantities of an inert gas. The gas was colorless, odorless, tasteless, insoluble and incombustible, so what could it be but nitrogen? He reported it as nitrogen.

When Ramsay finally came across this work some years later, he felt dissatisfied. A decision based on purely negative evidence seemed weak to him. He got hold of another uranium-containing mineral, collected the inert gas (which was there, sure enough), heated it and studied its spectrum. The lines were nothing like those of nitrogen. Instead, they were precisely those reported long ago by Janssen and Lockyer as having been found in sunlight. And so, in 1895, twenty-seven years after Lockyer's original assertion, the element of the sun was found on earth. Helium did exist and it was an element. Fortunately, Janssen and Lockyer lived to see themselves vindicated. Both lived well into their eighties, Janssen dying in 1907 and Lockyer in 1920.

Helium proved interesting at once. It was the lightest of the inert gases; lighter, in fact, than any known substance but hydrogen. The helium atom had only one layer of electrons and this innermost layer can only hold two electrons. Helium has those two electrons and is therefore inert; in fact, it is the most inert of all the inert gases, and therefore of all known substances.

This extreme inertness showed up almost at once in its liquefaction point; the temperature, that is, at which it could be turned into a liquid.

When neighboring atoms (or molecules) of a substance attract each other tightly, the substance hangs together all in a piece and is solid. It can be heated to a liquid and even to a gas, the transitions coming at those temperatures where the heat energy overcomes the attractive forces between the atoms or molecules. The weaker those attractive forces, the lower the temperature required to vaporize the substance.

If the attractive force between the atoms or molecules is low enough, so little heat is required to vaporize the substance that it remains gaseous at ordinary temperatures and even, sometimes, under conditions of great cold.

Particularly weak attractive forces exist when atoms or molecules have the stable eight-electron arrangement in their outermost electron shells. A nitrogen molecule is composed of two nitrogen atoms which have so arranged themselves that each owns at least a share in eight electrons in its outer shell. The same is true for other simple molecules, such as those of chlorine, oxygen, carbon monoxide, hydrogen and so on. All these are therefore gases that do not liquefy until very low temperatures are reached.

Little by little the chemists perfected their techniques for attaining low temperatures and liquefied one gas after another. The following table gives a measure of their progress, the liquefaction points being given in degrees Kelvin; that is, the number of Centigrade degrees above absolute zero.

Gas	Year first liquefied	Density (grams per liter)	Liquefaction point (° K.)
Chlorine	1805	3.214	239
Hydrogen bromide	1823	3.50	206
Ethylene	1845	1.245	169
Oxygen	1877	1.429	90
Carbon monoxide	1877	1.250	83
Nitrogen	1877	1.250	77
Hydrogen	1900	0.090	20

Now, throughout the 1870's and 1880's when low-temperature work was becoming really intense, it seemed quite plain that hydrogen was going to be the hardest nut of all to crack. In general, the liquefaction point went down with density, and hydrogen was by far the least dense of all known gases and should therefore have the lowest liquefaction point. Consequently, when hydrogen was conquered, the last frontier in this direction would have fallen.

And then, just a few years before hydrogen was conquered, it lost its significance, for the inert gases had been discovered. The electronically-satisfied atoms of the inert gases had so little attraction for each other that their liquefaction points were markedly lower than other gases of similar density. You can see this in the following table which includes all the inert gases but helium:

Inert gas	Density (grams per liter)	Liquefaction point (° K.)
Radon	9.73	211
Xenon	5.85	167
Krypton	3.71	120
Argon	1.78	87
Neon	0.90	27

As you see, radon, xenon and krypton, all denser than chlorine, have lower liquefaction points than that gas. Argon, denser than ethylene, has a markedly lower liquefaction point than that gas; and neon, ten times as dense as hydrogen, has almost as low a liquefaction point as that lightest of all gases.

The remaining inert gas, helium, which is only twice as dense as hydrogen, should, by all logic, be much more difficult to liquefy. And so it proved at once. At the temperature of liquid hydrogen, helium remained obstinately gaseous. Even when temperatures were dropped to the point where hydrogen solidified (13° K.), helium remained gaseous.

It was not until 1908 that helium was liquefied. The Dutch physicist Heike Kammerlingh Onnes turned the trick; the liquefaction of helium was found to take place at 4.2° K. By allowing liquid helium to evaporate under insulated conditions, Onnes chilled it further to 1° K.

Even at 1° K., there was no sign of solid helium, however. As a matter of fact, it is now established that helium never solidifies at ordinary pressures, not even at absolute zero, where all other known substances are solid. Helium (strange element) remains liquid. There is a reasonable explanation for this. Although it is usually stated that at absolute zero all atomic and molecular motions cease, quantum mechanics shows that there is a very small residual motion that never ceases. This bit of energy suffices to keep helium liquid. To be sure, at a temperature of 1° K. and the pressure of about 25 atmospheres, solid helium can be formed.

Liquid helium has something more curious to demonstrate than mere frigidity. When it is cooled below 2.2° K. there is a sudden change in its properties. For one thing, helium suddenly begins to conduct heat just about perfectly. In any ordinary liquid, within a few degrees of the boiling point, there are always localized hot spots where heat happens to accumulate faster than it can be conducted away. There bubbles of vapor appear, so that there is the familiar agitation one associates with boiling.

Helium above 2.2° K. ("helium I") also behaves like

this. Helium below 2.2° K. ("helium II"), however, vaporizes in absolute stillness, layers of atoms peeling off the top. Heat conduction is so nearly perfect that no part of the liquid can be significantly warmer than any other part and no bubbling takes place anywhere.

Furthermore, helium II has practically no viscosity. It will flow more easily than a gas and make its way through apertures that would stop a gas. It will form a layer over glass, creeping up the inner wall of a beaker and down the outer at a rate that makes it look as though it were pouring out of a hole in the beaker bottom. This phenomenon is called "superfluidity."

Odd properties are to be found in other elements at liquid helium temperatures. In 1911, Onnes, was testing the electrical resistance of mercury at the temperature of liquid helium. Resistance drops with temperature, and Onnes expected resistance to reach unprecedentedly low values; but he didn't expect it to disappear altogether. Yet it did. At a temperature of 4.12° K., the electrical resistance of mercury completely vanished. This is the phenomenon of "superconductivity."

Metals other than mercury can also be made superconductive. In fact, there are a few substances that can be superconductive at liquid hydrogen temperatures. Some niobium alloys become superconductive at temperatures as high as 18° K.

Superconductivity also involves an odd property with respect to a magnetic field. There are some substances that are "diamagnetic"; that is, which seem to repel magnetic lines of force. Fewer lines of force will pass through such substances than through an equivalent volume of vacuum. Well, any substance that is superconductive is completely diamagnetic; no lines of force enter it at all.

If the magnetic field is made strong enough, however, some lines of force eventually manage to penetrate the diamagnetic substance, and when that disruption of perfection takes place, all other perfections, including superconductivity, vanish. (It is odd to speak of perfection in nature. Usually perfections are the dreams of the theorist; the perfect gas, the perfect vacuum and so on. It is only at liquid helium temperatures that true perfection seems to enter the world of reality; hence the title of this chapter.)

The phenomenon of superconductivity has allowed the invention of a tiny device that can act as a switch. In simplest form, it consists of a small wire of tantalum wrapped about a wire of niobium. If the wires are dipped in liquid helium so that the niobium wire is superconductive, a tiny current passed through it will remain indefinitely, until another current is sent through the tantalum wire. The magnetic field set up in the second case, disrupts the superconductivity and stops the current in the niobium.

Properly manipulated, such a "cryotron" can be used to replace vacuum tubes or transistors. Tiny devices consisting of grouped wires, astutely arranged, can replace large numbers of bulky tubes or moderately bulky transistors, so that a giant computing machine of the future may well be desk-size or less if it is entirely "cryotronized."

The only catch is that for such a cryotronized computer to work, it must be dipped wholly into liquid helium. The liquid helium will be vaporizing continually so that each computer will, under these conditions, act as an eternally continuing drain on earth's helium supply.

Which brings up the question, of course, whether we have enough helium on earth to support a society in which helium-dipped computers are common.

The main, and, in fact, only commercial source of the inert gases other than helium is the atmosphere, which contains in parts per million, by weight:

Argon	12,800
Neon	12.5
Krypton	2.9
Helium	0.72
Xenon	0.36
Radon	(trace)

This means that the total atmospheric content of helium is 4,500,000,000 tons, which seems a nice tidy sum until you remember how extravagantly that weight of gas is diluted with oxygen and nitrogen. Helium can be obtained from liquid air, but only at tremendous expense.

(I would like to interrupt myself here to say that atmospheric helium consists almost entirely of the single isotope, helium-4. However, traces of the stable isotope, helium-3, are formed by the breakdown of radioactive hydrogen-3, which is, in turn, formed by the cosmic-ray bombardment of the atmosphere. Pure helium-3 has been studied and found to have a liquefaction point of only 3.2° K., a full degree lower than that of ordinary helium. However, helium-3, does not form the equivalent of the superfluid helium II. Only one atom of atmospheric helium per million is helium-3, so that the entire atmospheric supply amounts to only about 45,000 tons. Helium-3 is probably the rarest of all the stable isotopes here on earth.)

But helium, at least, is found in the soil as well as in the atmosphere. Uranium and thorium give off alpha particles, which are the nuclei of helium atoms. For billions of years, therefore, helium has been slowly collecting in the earth's crust (and remember that helium was first discovered on earth in uranium ore and not in the atmosphere). The earth's crust is estimated to contain helium to the extent of about 0.003 parts per million by weight. This means that the supply of helium in the crust is about twenty million times the supply in the atmosphere, but the dilution in the crust is nevertheless even greater than in the atmosphere.

However, helium is a gas. It collects in crevices and crannies and can come boiling up out of the earth under the right conditions. In the United States, particularly, wells of natural gas often carry helium to the extent of 1 percent, sometimes to an extent of up to 8 or even 10 percent.

But natural gas is a highly temporary resource which we are consuming rapidly. When the gas wells peter out, so will the helium, with the only remaining supply to be found in great dilution in the atmosphere or in even greater dilution in the soil.

It is possible to imagine, then, a computerized society of the future down to its last few million cubic feet of easily-obtainable helium. What next? Scrabble for the traces in air and soil? Make do with liquid hydrogen? Abandon cryotronized computers and try to return to the giant inefficient machines of the past? Allow the culture, completely dependent on computers, to collapse?

I've been thinking about that and here is the result of my thinking.

Such a threatened society ought to have developed space travel—why not?—so that they need not seek helium only here on earth.

Of course, the biggest source of helium in a solar system is the sun, but I see no way of snaking helium out of the sun in the foreseeable future.

The next biggest source of helium is Jupiter, which has an atmosphere that is probably thousands of miles deep, terrifically dense, and which is perhaps one-third helium by volume. Some recent theories suggest that the atmosphere is almost all helium. Milking Jupiter for helium doesn't sound easy, either, but it is conceivable.

Suppose mankind could establish a base on Jupiter V, Jupiter's innermost satellite. They would then be circling a mere 70,000 miles above the visible surface of Jupiter (which is actually the upper reaches of its atmosphere). Considerable quantities of helium-loaded gas must float even higher above Jupiter than that (and therefore closer to Jupiter V).

I can imagine a fleet of unmanned ships leaving Jupiter V in a probing orbit that will carry it down toward Jupiter's surface and back, collecting and compressing gas as it does so. Such gas will be easy to separate into its components; and the helium can be liquefied far more easily out on Jupiter V than here on earth, because the temperature is lower to begin with out there.

Uncounted tons of helium may prove easy to collect, liquefy and store. The next logical step would be to refrain from shipping that precious stuff anywhere else, even to earth. Why expend the energy and why undergo the tremendous losses that would be unavoidable in transit?

Instead, why not build the computers right there on Jupiter V?

And that is the vision I have, the one I mentioned at

the start of the chapter. It is the vision of Jupiter V—of all places—as the nerve center of the solar system. I see this small world, a hundred miles in diameter, extracting its needed helium from the bloated world it circles and slowly being converted into one large mass of interlocking computers, swimming in the most unusual liquid that ever existed.

However, I don't think I'll have the luck of Janssen and Lockyer. Call me a pessimist, if you wish, but somehow I don't think I'll live to see this.

6 The Weighting Game

Scientific theories have a tendency to fit the intellectual fashions of the time.

For instance, back in the fourth century B.C., two Greek philosophers, Leucippus of Miletus and Democritus of Abdera, worked out an atomic theory. All objects, they said, were made up of atoms. There were as many different kinds of atoms as there were fundamentally different substances in the universe. (The Greek recognized four fundamentally different substances, or "elements": fire, air, water and earth.)

By combinations of the elements in varying proportions, the many different substances with which we are familiar were formed. Through a process of separation and recombination in new proportions, one substance could be converted to another.

All this was very well, but in what fashion did these atoms differ from one another? How might the atom of one element be distinguished from the atom of another?

Now, mind you, atoms were far too small to see or to detect by any method. The Greek atomists were therefore perfectly at liberty to choose any form of distinction they wished. Perhaps different atoms had different colors, or different reflecting powers or bore little labels written in fine Attic Greek. Or perhaps they varied in hardness, in odor, or in temperature. Any of these might have sufficed as the basis for some coherent structural theory of the universe, given ingenuity enough—and if the Greeks had anything at all, it was ingenuity.

But here was where intellectual fashion came in. The Greek specialty was geometry. It was almost (though not quite) the whole of mathematics to them, and it invaded all other intellectual disciplines as far as possible. Consequently, if the question of atomic distinctions came up, the answer was inevitably geometric.

The atoms (the Greek atomists decided) differed in shape. The atoms of fire might be particularly jagged, which was why fire hurt. The atoms of water were smoothly spherical, perhaps, which was why water flowed so easily. The atoms of earth, on the other hand, could be cubical, which was why earth was so solid and stable. And so on.

This all had the merit of sounding very plausible and rational, but since there was no evidence for atoms at all, one way or the other, let alone atoms of different shapes, it remained just an intellectual exercise; not any more valid, necessarily, than the intellectual exercises of Greek philosophers who were non-atomist in their thinking. The non-atomists were more persuasive in their exercises, and atomism remained a minority view—very much in the minority —for over two thousand years.

Atomism was revived in the first decade of the nineteenth century by the English chemist John Dalton. He, too, believed everything was made up of atoms which combined and recombined in different proportions to make up all the substances we know.

In Dalton's time, the notion of elements had changed to the modern one, so that he could speak of atoms of fire and water. Furthermore, a vast array of chemical observations had been recorded during the seventeenth and eighteenth centuries, which could all be neatly explained by an atomic theory. This made the existence of atoms (still unseen and unseeable) a much more useful hypothesis than it had been in Greek times.

But now Dalton was faced with the same problem the Greeks had faced. How was an unseeable atom of one type to be distinguished from an unseeable atom of another type?

Well, the science of 1800 was no longer geometric. It was merely metric. That is, it was based on the measurement of three fundamental properties: mass (commonly miscalled "weight"), distance and time. These three together were sufficient to deal with the mechanical Newtonian universe.

Consequently, Dalton followed the intellectual fashion of the times and ignored form and shape. All atoms to him were featureless little spheres without internal structure. Instead, he automatically thought of mass, distance and time, and of these three the most clearly applicable was mass. As part of his theory, he therefore decided that atoms were distinguishable by mass only. All atoms of a particular element had identical mass, while the atoms of any one element had a mass that was different from those of any other element.

Dalton then went on—and this was perhaps his greatest single contribution—to try to determine what those different masses were.

Now there was no question of actually determining the mass of an atom in grams. That was not possible for quite some time to come. Relative masses, however, were another thing entirely.

For instance, hydrogen and oxygen atoms combine to form a molecule of water. (A "molecule" is the name applied to any reasonably stable combination of atoms.) It can be determined by analysis that in forming water, each gram of hydrogen combines with 8 grams of oxygen. In similar fashion, it could be determined that 1 gram of hydrogen always combines with just 3 grams of carbon to form methane. And, sure enough, 3 grams of carbon always combine with just 8 grams of oxygen to form carbon monoxide.

In this way, we are determining "equivalent weights"; that is, the weights of different elements that are equivalent to each other in the formation of compounds. (A "compound" is a substance whose molecules are made up of more than one kind of atom.) If we set the equivalent weight of hydrogen arbitrarily equal to 1, the equivalent weight of carbon is 3 and the equivalent weight of oxygen is 8.

How can this be related to atoms? Well, Dalton made the simplest assumption (which is what one should always do) and decided that one atom of hydrogen combined with one atom of oxygen to form water. If that were so, then the atoms of oxygen must be eight times as heavy as the atoms of hydrogen, which is the best way of explaining why 1 gram of hydrogen combines with 8 grams of oxygen. The same number of atoms are on each side, you see, but the oxygen atoms are each eight times as heavy as the hydrogen atoms.

Thus, if we arbitrarily set the atomic weight of hydrogen equal to 1, then the atomic weight of oxygen is 8. By the same reasoning, the atomic weight of carbon is 3, if a molecule of methane consists of one atom of carbon combined with one atom of hydrogen.

The inevitable next question, though, is just this: How valid is Dalton's assumption? Do atoms necessarily combine one-to-one? The answer is: No, not necessarily.

Whereas 3 grams of carbon combine with 8 grams of oxygen to form carbon monoxide, 3 grams of carbon will also combine with 16 grams of oxygen to form carbon dioxide.

Well, then, if we assume that the carbon monoxide molecule is made up of one atom of carbon and one atom of oxygen, and let C represent carbon and O represent oxygen, we can write the molecule of carbon monoxide as CO. But if carbon combines with twice the quantity of oxygen to form a molecule of a substance with different properties, we can assume, on the basis of atomic theory, that each atom of carbon combines with *two* atoms of oxygen to form carbon dioxide. The formula of carbon dioxide is therefore CO_2.

On the other hand, if you reasoned that the molecule of carbon dioxide was CO, then the molecule of carbon monoxide would have to be C_2O. The first alternative, presented in the previous paragraph, happens to be the correct one, but in either alternative we come up against a molecule in which one atom of one element combines with two atoms of another element.

Once you admit that a molecule may contain more than one of a particular kind of atom, you must re-examine the structure of the water molecule. Must it be formed of an atom of hydrogen and one of oxygen, with a formula of HO? What if its formula were HO_2 or HO_4 or H_4O, or, for that matter, $H_{17}O_{47}$?

Fortunately, there was a way of deciding the matter. In 1800, two English chemists, William Nicholson and Anthony Carlisle, had shown that if an electric current were passed through water, hydrogen and oxygen gases were produced. It was quickly found that hydrogen was produced in just twice the volume that oxygen was. Thus, although the ratio of hydrogen to oxygen in water was 1 to 8 in terms of mass, it was 2 to 1 in terms of volume.

Was there any significance to this? Perhaps not. The atoms in hydrogen gas might be spaced twice as far apart as the atoms in oxygen gas, so that the volume difference might have no relation to the number of atoms produced.

However, in 1811, an Italian chemist, Amedeo Avogadro, suggested that in order to explain the known behavior of gases in forming chemical combinations, it was necessary to assume that equal volumes of different gases contained equal numbers of particles. (The particles could be either atoms or molecules.)

Therefore, if the volume of hydrogen produced by the electrolysis of water was twice the volume of oxygen, then twice as many particles of hydrogen were produced as of oxygen. If these particles are assumed to be atoms, or molecules containing the same number of atoms in both hydrogen or oxygen (the latter turned out to be true), then the water molecule contained twice as many hydrogen atoms as oxygen atoms.

The formula for water could not be HO, therefore, but had to be, at the very simplest, H_2O. If 8 grams of oxygen combined with 1 gram of hydrogen, it meant that the single oxygen atom is eight times as heavy as the two hydrogen atoms taken together. If you still set the atomic weight of hydrogen at 1, then the atomic weight of oxygen is equal to 16.

In the same way, it was found eventually that the formula of methane was CH_4, so that the one carbon atom had to be three times as heavy as the four hydrogen atoms taken together. (The equivalent weight of carbon is 3, remember.) Thus, if the atomic weight of hydrogen is 1, then the atomic weight of carbon is 12.

"Avogadro's hypothesis," as it came to be called, made it possible to come to another decision. One liter of hydrogen combined with one liter of chlorine to form hydrogen chloride. It was therefore a fair working assumption to suppose that the hydrogen chloride molecule was made up of one atom of hydrogen and one atom of chlorine. The formula of hydrogen chloride (allowing "Cl" to symbolize "chlorine") could then be written HCl.

The liter of hydrogen and the liter of chlorine contain equal numbers of particles, Avogadro's hypothesis tells us. If we assume that the particles consist of individual atoms, then the number of hydrogen chloride molecules formed must be only half as many as the total number of hydrogen atoms and chlorine atoms with which we start. (Just as the number of married couples is only half as many as the total number of men and women, assuming everyone is married.)

It should follow that the hydrogen chloride gas that is formed has only half the total volume of the hydrogen and chlorine with which we start. One liter of hydrogen plus one liter of chlorine (two liters in all) should produce but one liter of hydrogen chloride.

However, this is not what happens. A liter of hydrogen and a liter of chlorine combine to form *two* liters of hydrogen chloride. The total volume of gas does not change and therefore the total number of particles cannot change. The simplest way out of the dilemma is to assume that hydrogen gas and chlorine gas are not collections of single atoms after all, but collections of molecules, each of which is made up of two atoms.

One hydrogen molecule (H_2) would combine with one chlorine molecule (Cl_2) to form two molecules of hydrogen chloride (HCl, HCl). The total number of particles would not change and neither would the total volume. By similar methods it could be shown that oxygen gas is also made up of molecules containing two atoms apiece (O_2).

Using this sort of reasoning, plus other generalizations I am not mentioning, it was possible to work out atomic weights and molecular structures for a whole series of substances. The one who was busiest at it was a Swedish chemist named Jöns Jakob Berzelius who, by 1828, had put a series of atomic weights that were pretty darned good even by modern standards.

However, the course of true love never does run smooth; nor, it seems, does the course of science. A chemist is as easily confused as the next guy; and all during the first half of the nineteenth century, the words "atom" and "molecule" were used interchangeably. Few chemists got them straight, and few distinguished the atomic weight of chlorine, which was 35.5, from the molecular weight of chlorine, which was 71 (since a molecule of chlorine contains two atoms). Then again, the chemists confused atomic weight and equivalent weight, and had difficulty seeing that though the equivalent weights of carbon and oxygen were 3 and 8 respectively, the atomic weights were 12 and 16 respectively. (And, to make matters worse, the molecular weight of oxygen was 32.)

This reduced all chemical calculations, upon which decisions as to molecular structure were based, to sheer chaos. Things weren't too bad with the simple molecules of inorganic chemistry, but in organic chemistry, where molecules contained dozens of atoms, the confusion was ruinous. Nineteen different formulas were suggested for acetic acid, which, with a molecule containing merely eight atoms, was one of the simplest of the organic compounds.

Then, in 1860, a German chemist named Friedrich August Kekule organized the first International Chemical Congress in order to deal with the matter. It assembled at Karlsruhe in Germany.

The hit of the Congress was an Italian chemist named Stanislao Cannizzaro. In formal speeches and in informal talks he hammered away at the importance of straightening out the matter of atomic weights. He pointed out how necessary it was to distinguish between atoms and molecules and between equivalent weights and atomic weights. Most of all, he explained over and over again the significance of the hypothesis of his countryman, Avogadro, a hypothesis most chemists had been ignoring for half a century.

He made his case; and over the course of the next decade, chemistry began to straighten up and fly right.

The result was pure gold. Once Cannizzaro had sold the notion of atomic weights, a few chemists began to arrange the elements in the order of increasing atomic weight to see what would happen. About sixty elements were known in 1860, you see, and they were a bewildering variety of types, makes and models. No one could predict how many more elements remained to be found nor what their properties might be.

The first attempts to make an atomic weight arrangement seemed to be interesting, but chemists as a whole remained unconvinced that it was anything more than a form of chemical numerology. Then along came a Russian chemist named Dmitri Ivanovich Mendeleev who, in 1869, made the most elaborate arrangement yet. In order to make his table come out well, he left gaps which, he insisted, signified the presence of yet undiscovered elements. He predicted the properties of three elements in particular. Within a dozen years those three elements were discovered, and their properties jibed in every particular with those predicted by Mendeleev.

The sensation was indescribable. Atomic weights were the smash of the season and a number of chemists began to devote their careers to the more-and-more accurate determination of atomic weights. A Belgian chemist, Jean Servais Stas, had already produced a table of atomic weights far better than that of Berzelius in the 1860's, but the matter reached its chemical peak in the first decade of the twentieth century, just a hundred years after Dalton's first attempts in this direction. The American chemist Theodore William Richards analyzed compounds with fantastic precautions against impurities and error and obtained such accurate atomic weight values that he received the 1914 Nobel Prize in chemistry for his work.

But as fate would have it, by that time atomic weights had gotten away from chemists and entered the domain of the physicist.

The break came with the discovery of the subatomic particles in the 1890's. The atom was *not* a featureless particle, it turned out. It was a conglomerate of still smaller particles, some of which are electrically charged.

It turned out then that the fundamental distinction between atoms of different elements was not the atomic weight at all, but the quantity of positive electric charge upon the nucleus of the atom. (Again, this fitted the intellectual fashion of the times, for as the nineteenth century wore on, the mechanical Newtonian universe gave way to a universe of force fields according to the theories of the English chemist Michael Faraday and the Scottish physicist James Clerk Maxwell. Electric charge fits into this force field scheme.)

It turned out that most elements consisted of varieties of atoms of somewhat different atomic weight. These varieties are called "isotopes"; for a discussion of them, see Chapter 7.

What we have been calling the atomic weight is only the average of the weights of the various isotopes making up the element.

Physicists began to determine the relative masses of the individual isotopes by nonchemical methods, with a degree of accuracy far beyond the ordinary chemical methods even of Nobel laureate Richards. To get an accurate atomic weight it was then only necessary to take a weighted average of the masses of the isotopes making up the elements, allowing for the natural percentage of each isotope in the element as found in nature.

The fact that atomic weights had thus become a physical rather than a chemical measurement might not have been embarrassing, even for the most sensitive chemist, were it not for the fact that physicists began to use atomic weight values slightly different from that used by the chemists. And what made it really bad was that the physicists were right and the chemists wrong.

Let me explain.

From the very beginning, the measurement of atomic weights had required the establishment of a standard. The most logical standard seemed to be that of setting the atomic weight of hydrogen equal to 1. It was suspected then (and it is known now) that hydrogen possessed the lightest possible atom, so setting it equal to 1 was the most natural thing in the world.

The trouble was that in determining atomic weights one had to start with equivalent weights. (In the beginning, anyway.) To determine equivalent weights, one needed to work with two elements that combined easily. Now hydrogen combined directly with but few elements, whereas oxygen combined directly with many. It was a matter of practical convenience to use oxygen, rather than hydrogen, as a standard.

This made a slight modification necessary.

Atomic weights, after all, don't match in exact whole-number ratios. If the atomic weight of hydrogen is set at exactly 1, then the atomic weight of oxygen is not quite 16. It is, instead, closer to 15.9. But if oxygen is the element most often used in calculating equivalent weights, it would be inconvenient to be forever using a figure like 15.9. It it an easy alternative to set the atomic weight of oxygen exactly at 16 and let the atomic weight of hydrogen come out a trifle over 1. It comes out to 1.008, in fact.

We can call this the "O = 16" standard. It made chemists very happy, and there arose nothing to challenge it until the 1920's. Then came trouble.

Oxygen, it was discovered in 1929, was a mixture of three different isotopes. Out of every 100,000 oxygen atoms, 99,759, to be sure, had an atomic weight of about 16. Another 204, however, had an atomic weight of about 18, while the remaining 37 had an atomic weight of 17. (The isotopes can be symbolized as O^{16}, O^{17} and O^{18}.)

This meant that when chemists set oxygen equal to 16, they were setting a weighted average of the three isotopes equal to 16. The common oxygen isotope was just a little under 16 (15.9956, to be exact), and the masses of the relatively few oxygen atoms of the heavier isotopes pulled that figure up to the 16 mark.

Physicists working with individual nuclei were more interested in a particular isotope than in the arbitrary collection of isotopes in an element. In this they had logic on their side, for the mass of an individual isotope is, as far as we know, absolutely constant, while the average mass of the atoms of an element fluctuates slightly as the mixture varies a tiny bit under different conditions.

Now we have two scales. First there is the "chemical atomic weight" on the "O = 16" standard. Second, there is the "physical atomic weight" on the "O^{16} = 16" standard.

On the chemical atomic weight scale, the atomic weight of oxygen is 16.0000; while on the physical atomic weight scale, the heavier oxygen isotopes pull the average weight up to 16.0044. Naturally, all the other atomic weights must change in proportion, and every element has an atomic weight that is 0.027 per cent higher on the physical scale than on the chemical scale. Thus, hydrogen has a chemical atomic weight of 1.0080 and a physical atomic weight of 1.0083.

This isn't much of a difference, but it isn't neat. Chemists and physicists shouldn't disagree like that. Yet chemists, despite the weight of logic against them, were reluctant to abandon their old figures and introduce confusion when so many reams of calculations in the chemical literature had been based on the old chemical atomic weights.

Fortunately, after three decades of disagreement, a successful compromise was reached.

It occurred to the physicists that, in using an "$O^{16} = 16$" standard, they were kowtowing to a chemical prejudice which no longer had validity. The only reason that oxygen was used as the standard was the ease with which oxygen could be used in determining equivalent weights.

But the physicists weren't using equivalent weights; they didn't give a continental for equivalent weights. They were determining the masses of charged atoms by sending them through a magnetic field of known strength and measuring the effect upon the path of those atoms.

In this connection, oxygen atoms were not the best atoms to use as standard; carbon atoms were. The mass of the most common carbon isotope, C^{12}, was more accurately known than that of any other isotope. Moreover, C^{12} had a mass that was 12.003803 on the physical scale but was almost exactly 12 on the chemical scale.

Why not, then, set up a "$C^{12} = 12$" scale? It would be just as logical as the "$O^{16} = 16$" scale. What's more, the "$C^{12} = 12$" scale would be almost exactly like that of the chemical "$O = 16$" scale.

In 1961, the International Union of Pure and Applied Physics issued a ukase that this be done. The mass of C^{12}

was set at exactly 12.000000, a decrease of 0.033 percent. Naturally, the masses of all other isotopes had to decrease by exactly the same percentage, and the "$C^{12} = 12$" scale fell very slightly below the "$O = 16$" scale.

Thus, the chemical atomic weights of hydrogen and oxygen were 1.0080 and 16.0000 respectively. The atomic weights on the new physical scale were 1.00797 and 15.9994 respectively.

The difference is now only 0.003 per cent, only one-tenth of the difference between the chemical scale and the old physical scale.

The chemists could no longer resist; the difference was so small that it would not affect any of the calculations in the literature. Consequently, the International Union of Pure and Applied Chemistry has also accepted the "$C^{12} = 12$" scale. Physicists and chemists once again, as of 1961, speak the same atomic weigh language.

Note, too, how it was done. The physicists made most of the adjustment in actual value and that was a victory for the chemists. The chemists, on the other hand, adopted the logic of the single isotope as standard, and that was a victory for the physicists.

And since the standard which has now been adopted is the most accurate one yet, the net result is a victory for everybody.

Now *that* is the way to run the world—but I'll refrain from trying to point a moral.

7 The Evens Have It

Some time ago, I was asked (by phone) to write an article on the use of radioisotopes in industry. The gentleman doing the asking waxed enthusiastic on the importance of isotopes, but after a while I could stand it no more, for he kept pronouncing it ISS-o-topes, with a very short " i."

Finally, I said, in the most diffident manner I could muster, "EYE-so-topes, sir," giving it a very long "i."

"No, no," he said impatiently, "I'm talking about ISS-o-topes."

And so he did, to the very end, and on subsequent phone calls too. But I fooled him. I eventually wrote the article about EYE-so-topes.

Yet it left a sore spot, for having agreed to do the article, I was forced to deal with only the practical applications of isotopes, a necessity which saddened me. There is much that is impractical about isotopes that I would like to discuss, and I will do so here.

The way in which "isotope" came into the scientific vocabulary is a little involved. After two millennia of efforts, most of the elements making up the universe had been isolated and identified. In 1869, the Russian chemist Dmitri Ivanovich Mendeleev arranged the known elements in order of atomic weights and showed that a table could be prepared in which the elements, in this order, could be so placed as to make those with similar properties fall into neat columns.

By 1900, this "periodic table" was a deified adjunct of chemistry. Each element had its place in the table; almost each place had its element. To be sure, there were a few places without elements; but that bothered no one since everyone knew that the list of know elements was incomplete. Eventually, chemists felt certain, an element would be discovered for every empty place in the table. And they were right. The last hole was filled in 1948, and additional elements were discovered beyond the last known to Mendeleev. As of now, 103 different elements are known.

After 1900, however, the much more serious converse of the situation arose. Substances were found among the radioactive breakdown products of uranium and thorium which had to be classified as new elements by nineteenth-century standards, since they had properties unlike those of any other elements—and yet there was no place for them in the periodic table.

Eventually several scientists, notably the British physicist Frederick Soddy, swallowed hard and decided that it was possible for two or more elements to occupy the same place in the periodic table. In 1913, Soddy suggested the name "isotope" for such elements, from Greek words meaning "same place."

An explanation rehabilitating the periodic table followed in due course. The New Zealand-born British physicist Ernest Rutherford had already (in 1906) shown that the atom consisted of a tiny central nucleus containing positively-charged protons and of a comparatively vast outer region in which negatively-charged electrons whirled. The number of protons at the center is equal to the number of electrons in the outskirts, and since the size of the positive electric charge on a proton (arbitrarily set at $+1$) is exactly equal to the size of the negative electric charge on an electron (which is naturally, -1), the atom as a whole is electrically neutral.

The next step was taken by a young English physicist named Henry Gwyn-Jeffreys Moseley. By studying the wave lengths of the X rays emitted, under certain conditions, by various elements, he was able to deduce that the total positive charge on the nucleus of each element had a characteristic value. This is called the "atomic number."

For instance, the chromium atom has a nucleus with a positive charge of 24, the manganese atom one of 25 and the iron atom one of 26. We can say then that the atomic numbers of these elements are 24, 25, and 26 respectively. Furthermore, since the positive charge is entirely due to the proton content of the nucleus, we can say that these three elements have 24, 25, and 26 protons in their nuclei, respectively, and that circling these nuclei are 24, 25, and 26 electrons, respectively.

Now throughout the nineteenth century it had been held that all atoms of an element were identical. This was only an assumption, but it was the easiest way of explaining the fact that all samples of an element had identical chemical properties and identical atomic weights.

But this was when atoms were viewed as hard, indivisible, featureless spheres. How did the situation stand up against the twentieth-century notion that the atoms were complex collections of smaller particles?

X-ray data showed that the atomic number of an element was a matter of absolute uniformity. All atoms of a particular element had the same number of protons in the nucleus and therefore the same number of electrons in the outskirts. Through the 1920's it was shown that the chemical properties of a particular element depended on the number of electrons it contained and that therefore all atoms of an element had identical chemical properties. Very good, so far.

The matter of atomic weight was not so straightforward. To begin with, it was known from the first days of nuclear-atomic theory that the nucleus must contain something other than protons. For instance: the nucleus of the hydrogen atom was the lightest known and it had a positive charge of 1. Consequently it seemed quite reasonable, and even inevitable, to suppose that the hydrogen nucleus was made up of a single proton. Its atomic weight, which had been set equal to 1 (not quite, but just about) long before the days when atomic structure had been worked out, turned out to make sense.

Helium, on the other hand, had an atomic weight of 4. That is, its nucleus was known to be four times as massive as the hydrogen nucleus. The natural conclusion seemed to be that it must contain four protons. However, its atomic number, representing the positive charge of its nucleus, was only 2. An equally natural conclusion from that seemed to be that the nucleus must contain only two protons.

With two different but natural conclusions, something had to be done. The only other subatomic particle known in the first decades of the twentieth century was the electron. Suppose then that the helium nucleus contained four protons and two electrons. The atomic weight would be 4 because the electrons weigh practically nothing. The atomic number, however, would be 2 because the charge on two of the protons would be cancelled by the charge on the two electrons.

There were difficulties in this picture of the nucleus, however. For instance, it gave the helium nucleus six separate particles, four protons and two electrons, and that didn't fit in with certain other data that were being accumulated. Physicists went about biting their nails and talking in low, glum voices.

Then, in 1932, the neutron was discovered by the English physicist James Charwick, and it turned out that all was right with the theory, after all. The neutron is equal in mass to the proton (just about), but has no charge at all. Now the helium nucleus could be viewed as consisting of two protons and two neutrons, you see. The positive charge and hence the atomic number would be 2 and the atomic weight would be 4. This would involve a total number of four particles in the helium nucleus, and that fit all data.

Now, how does the presence of neutrons in the nucleus of an atom affect the chemical properties? Answer: It doesn't; at least, not noticeably.

Take as an example the copper atom. It has an atomic number of 29, so every copper atom has twenty-nine protons in the nucleus and twenty-nine electrons in the outer reaches. But copper has an atomic weight of (roughly) 63, so the nucleus of the copper atom must contain, in addition to twenty-nine protons, thirty-four neutrons as well. The neutrons have no charge; they do not need to be balanced. The twenty-nine electrons balance the twenty-nine protons and, as far as they are concerned, the neutrons can go jump in the lake.

Well, then, suppose just for fun that a copper atom

happened to exist with a nucleus containing twenty-nine protons and thirty-six neutrons, two more neutrons, that is, than the number suggested in the previous paragraph. Such a nucleus would still require only twenty-nine electrons to balance the nuclear charge; and the chemical properties, which depend on the electrons only, would remain the same.

In other words, if we judge by chemical properties alone, the atoms of an element need *not* be identical. The number of neutrons in the nucleus could vary all over the lot and this would make no difference chemically. Since the periodic table points out chemical similarities and since the elements are defined by their chemical properties, it means that each place in the periodic table is capable of holding a large variety of different atoms, with different numbers of neutrons, *provided* the number of protons in all those atoms is held constant.

But how does this affect the atomic weight?

The two varieties of copper atoms would, naturally, be well mixed at all times. Why not? Since they would have identical chemical properties, they would travel the same path in geochemical processes; all of them would react equally with the environment about them, go into solution and out of solution at the same time and to the same extent. They would be inseparable; in the end, any sample of an element found in nature, or prepared in the laboratory, would contain the same even mixture of the two copper isotopes.

In obtaining the atomic weight of an element, then, nineteenth-century chemists were getting the *average* weight of the atoms of that element. The average would always be the same (for anything *they* could do), but that did *not* mean that all the atoms were individually identical.

Then what happened to upset this comfortable picture once radioactivity was discovered?

Well, radioactive breakdown is a *nuclear* process, and whether it takes place or not, and how quickly, and in what fashion, depends on the arrangement of particles in the nucleus and has nothing to do with the electrons outside the nucleus. It follows that two atoms with nuclei containing the same number of protons but different numbers of neutrons would have identical chemical properties but different nuclear properties. It was the identical chemical properties that place them in the same spot in the periodic table. The different nuclear properties had nothing to do with the periodic table.

But in the first decade of the twentieth century, when the distinction between nuclear properties and chemical properties had not yet been made, there was this period of panic when it seemed that the periodic table would go crashing.

It was easy to distinguish between two isotopes (which, you now see, are two atoms with equal numbers of protons in their nuclei but different numbers of neutrons) if radioactivity was involved. What, however, if neither of two isotopes is radioactive? Is it even possible for there to be more than one non-radioactive isotope of a given element?

Well, if a plurality of non-radioactive isotopes of an element existed, they would differ in mass. A copper atom with 29 protons and 34 neutrons would have a "mass number" of 63, while one with 29 protons and 36 neutrons would have one of 65. (The expression "atomic weight" is reserved for the average masses of naturally occurring mixtures of isotopes of a particular element.)

In 1919, the English physicist Francis William Aston invented the mass spectrograph in which atoms in ionic form (that is, with one or more electrons knocked off so that each atom has a net positive charge) could be driven through a magnetic field. The ions follow a curved path in so doing, the sharpness of the curve depending on the mass of the particular ion. Isotopes having different masses end on different spots of a photographic plate, and from the intensities of darkening, the relative quantities of the individual isotopes can be determined. For instance, the 34-neutron copper atom makes up 70 per cent of all copper atoms while the 36-neutron copper atom makes up the remaining 30 per cent. This accounts for the fact that the atomic weight of copper is not exactly 63, but is actually 63.54.

To distinguish isotopes, chemists make use of mass

numbers. A copper atom with 29 protons and 34 neutrons has a mass number of 29 plus 34, or 63, and can therefore be referred to as "copper-63," while one with 29 protons and 36 neutrons would be "copper-65." In written form, chemical symbols plus superscriptions are used, as Cu^{63} and Cu^{65}.

By this system, only the total number of protons plus neutrons are given. Chemists shrug this off. They know the atomic number of each element by heart (or they can look it up when no one's watching them), and that gives them the number of protons in the nucleus. By subtracting the atomic number from the mass number, they get the number of neutrons.

But for our purposes, I am going to write isotopes with proton and neutron numbers both clearly stated, thus: copper-29/34 and copper-29/36. If I want to refer to both of them, I will write: copper-29/34,36. Fair enough?

With this background, we can now look at the isotopes more closely. For instance, we can divide them into three varieties. First, there are the radioactive ones that break down so rapidly (lasting no longer than a few million years at most) that any which exist now have arisen in the comparatively near past as a result of some nuclear reaction, either in nature or in the laboratory. I will call these the "unstable" isotopes. Although over a thousand of these are known, each one exists in such fantastically small traces (if at all) that they make themselves known only to the nuclear physicist and his instruments.

Secondly, there are isotopes which are radioactive but which break down so slowly (in hundreds of millions of years at the very least) that those which exist today have existed at least since the original formation of the earth. Each of them, despite its continuous breakdown, exists in nature in quantities that would make it detectable by old-fashioned, nineteenth-century chemical methods. I will call these the "semi-stable" isotopes.

Finally there are the isotopes which are not at all radioactive or are so feebly radioactive that even our most sensitive instruments cannot detect it. These are the "stable" isotopes.

In this chapter I shall concern myself only with the semi-stable and stable isotopes.

No less than 20 of the 103 elements known today possess only unstable isotopes and therefore exist in nature either in insignificant traces or not at all. These are listed in Table 1. Notice that all but two of these elements exist at the very end of the known list of elements, with atomic numbers running from 84 to 103. The only elements not on the list, within that stretch, are elements number 90 (thorium) and 92 (uranium), both of which, you will note, have an even atomic number. On the other hand, there are two elements in the list with atomic numbers below that range, elements number 43 (technetium) and 61 (promethium), both with odd atomic numbers.

TABLE 1—ELEMENTS WITHOUT STABLE OR
SEMI-STABLE ISOTOPES

Element	Atomic Number	Element	Atomic Number
Technetium	43	Plutonium	94
Promethium	61	Americium	95
Polonium	84	Curium	96
Astatine	85	Berkelium	97
Radon	86	Californium	98
Francium	87	Einsteinium	99
Radium	88	Fermium	100
Actinium	89	Mendelevium	101
Protactinium	91	Nobelium	102
Neptunium	93	Lawrencium	103

This means that there are exactly 83 elements which possess at least one stable or semi-stable isotope and which therefore occur in reasonable quantities on earth. (There is no stable or semi-stable isotope that does not occur in nature in reasonable quantities.) Some of these elements possess only one such isotope, some two, some three, and some more than three.

Now, it is an odd thing that although every chemistry

textbook I have ever seen always lists the elements, no book I have ever seen lists the isotopes in any systematic way.

For instance, I have never seen *anywhere* a complete list of all those elements possessing but a single stable or semi-stable isotope. I have prepared such a list (Table 2).

TABLE 2—ELEMENTS WITH ONE STABLE OR SEMI-STABLE ISOTOPE

Element	Proton/ Neutron	Element	Proton/ Neutron
Beryllium	4/5	Rhodium	45/58
Fluorine	9/10	Iodine	53/74
Sodium	11/12	Cesium	55/78
Aluminum	13/14	Praseodymium	59/82
Phosphorus	15/16	Terbium	65/94
Scandium	21/24	Holmium	67/98
Manganese	25/30	Thulium	69/100
Cobalt	27/32	Gold	79/118
Arsenic	33/42	Bismuth	83/126
Yttrium	39/50	Thorium	90/142*
Niobium	41/52		

*semi-stable

There are twenty-one elements with one stable or semi-stable isotope apiece, and you will notice that in every case but two (beryllium and thorium, first and last in the list) the solo isotopes have an odd number of protons in the nucleus and an even number of neutrons. These are the "odd/even" isotopes.

Let's next list the elements that possess two stable or semi-stable isotopes (Table 3). This list includes twenty-three elements, of which twenty possess odd numbers of protons.

If you look at the first three tables, you will see that of the 52 known elements with odd atomic numbers, 12 possess no stable or semi-stable isotopes, 19 possess just one stable or semi-stable isotope and 20 possess just two. The total comes to 51.

TABLE 3—ELEMENTS WITH TWO STABLE OR SEMI-STABLE ISOTOPES

Element	Proton/ Neutrons	Element	Proton/ Neturons
Hydrogen	1/0,1	Silver	47/60,62
Helium	2/1,2	Indium	49/64,66*
Lithium	3/3,4	Antimony	51/70,72
Boron	5/5,6	Lanthanum	57/61*,62
Carbon	6/6,7	Europium	63/88,90
Nitrogen	7/7,8	Lutetium	71/104,105*
Chlorine	17/18,20	Tantalum	73/107*,108
Vanadium	23/27*,28	Rhenium	75/110,112*
Copper	29/34,36	Iridium	77/114,116
Gallium	31/38,40	Thallium	81/122,124
Bromine	35/44,46	Uranium	92/143*,146*
Rubidium	37/48,50*		

*semi-stable

There is one and only one element of odd atomic number left unaccounted for; and if you follow down the lists, the missing element turns out to be number 19, which is potassium. Potassium has three stable or semi-stable isotopes, and I'll list it here without giving it the dignity of a table all to itself: Potassium-19/20,21*,22. (The asterisk after the neutron number denotes a semi-stable isotope.)

Of these, the semi-stable potassium-19/21* (the lightest of all the semi-stable isotopes) makes up only one atom in every ten thousand of potassium, so that this element just *barely* has more than two isotopes.

The 52 elements with odd atomic number contain, all told, 62 different stable or semi-stable isotopes. Of these, 53 contain an even number of neutrons, so that there are 53 odd/even stable or semi-stable isotopes in existence.

These can be broken up into 50 stable and 3 semi-stable (rubidium-37/50*, indium-49/66*, and rhenium-75/112*).

There are only nine stable or semi-stable isotopes of the atoms with odd atomic number that possess an odd number of neutrons as well. Table 4 contains a complete list of all the "odd/odd" isotopes, which are stable or semi-stable, in existence.

TABLE 4—THE STABLE OR SEMI-STABLE ODD/ODD ISOTOPES

Element	Proton/Neutron
Hydrogen	1/1
Lithium	3/3
Boron	5/5
Nitrogen	7/7
Potassium	19/21*
Vanadium	23/27*
Lanthanum	57/61*
Lutetium	71/105*
Tantalum	73/107*

*semi-stable

As you see, of these 9, fully 5 are semi-stable. This means that only 4 completely stable odd/odd isotopes exist in the universe. Of these, the odd/odd hydrogen-1/1 is outnumbered by the odd/even hydrogen-1/0 (I am calling zero an even number, if you don't mind) ten thousand to one. The odd/odd lithium-3/3 is outnumbered by the odd/even lithium-3/4 by thirteen to one, and the odd/odd boron-5/5 is outnumbered by the odd/even boron-5/6 by four to one. So three of the four stable odd/odd isotopes form minorities within their own elements.

This leaves nitrogen-7/7, an odd/odd isotope which is not only completely stable but which makes up 99.635 percent of all nitrogen atoms. It is in this respect, the oddest of all the odd/odds.

What about the elements of even atomic number?

There the situation is reversed. Only eight of the elements of even atomic number have no stable or semi-stable isotopes, and all of these are in the region beyond atomic number 83 where no fully stable and almost no semi-stable isotopes exist. What's more, the three semi-stable isotopes that do exist in that region all belong to elements of even atomic number.

There are two other elements of even atomic number with but a single stable or semi-stable isotope and three with but two stable isotopes. You pick all these up in the tables already presented .

This leaves 39 of the 51 elements of even atomic number, all possessing more than two stable isotopes. One of them, tin, possesses no less than ten stable isotopes. I will not tabulate these elements in detail.

Instead, I will point out that there are two varieties of isotopes where elements of even atomic number are involved. There are isotopes with odd numbers of neutrons ("even/odd") and those with even numbers ("even/even").

The data on stable and semi-stable isotopes are summarized in Table 5.

In sheer numbers of isotopes, the even/even group is preponderant, making up 60 percent of the total. The preponderance is even greater in mass.

TABLE 5—VARIETIES OF ISOTOPES

	Stable	Semi-Stable	Total
Even/Even	164	3	167
Even/Odd	55	2	57
Odd/Even	50	3	53
Odd/Odd	4	5	9
Total	273	13	
		Grand Total	286

Among the 43 elements of even atomic number that possess stable or semi-stable isotopes, only one lacks an even/even isotope. That is beryllium, with but one stable

99

or semi-stable isotope, beryllium-4/5, which is even/odd.

Of the 42 others, there is not one case in which the even/even isotopes do not make up most of the atoms. The even/odd isotope which is most common within its own element is platinum-78/117, which makes up one-third of all platinum atoms. Where an element of even atomic number has more than one even/odd isotope (tin has three), all of them together sometimes do even better. The record is the case of xenon-54/75 and xenon-54/77, which together make up almost 48 percent of all xenon atoms. In no case do the even/odd isotopes top the 50 per cent mark, except in the case of beryllium, of course.

What's more, the even/odd isotopes do best just in those elements which are least common. Platinum and zenon are among the rarest of all the elements with stable or semi-stable isotopes. It is precisely in the most common elements that the even/even isotopes are most predominant.

This shows up when we consider the structure of the earth's crust. I once worked out its composition in isotope varieties and this is the result:

even/even — 85.63 per cent
odd/even — 13.11 per cent
even/odd — 1.25 per cent
odd/odd — 0.01 per cent

Almost 87 per cent of the earth's crust is made up of the elements with even atomic numbers. And if the entire earth is considered, the situation is even more extreme. In Chapter 14, I will point out that six elements make up 98 per cent of the globe, these being iron, oxygen, magnesium, silicon, sulfur and nickel. Every one of these is an element of even atomic number. I estimate that the globe we live on is 96 per cent even/even.

Which is a shame, in a way. As a long-time science-fiction enthusiast and practicing nonconformist, I have always had a sneaking sympathy for the odd/odd.

III PHYSICS

8 Now Hear This!

The ancient Greeks weren't always wrong.

I am taking the trouble to say this strictly for my own good, for when I trace back the history of some scientific concept, I generally start with the Greeks, then go to great pains to show how their wrong guesses had to be slowly and painfully corrected by the great scientists of the sixteenth and seventeenth centuries, usually against the strenuous opposition of traditionalists. By the time I have done this on several dozen different occasions, I begin, as a matter of autohypnosis, to think that the only function served by the ancient philosophers was to put everyone on the wrong track.

And yet, not entirely so. In some respects we are still barely catching up to the Greeks. For instance, the Navy is now studying dolphins and porpoises. (These are small-sized relatives of the whale, which differ among themselves most noticeably in that dolphins have lips that protrude in a kind of beak while porpoises do not. I will, however, use the two terms interchangeably and without any attempt at consistency.)

In recent years, biologists have begun making observations concerning the great intelligence of these creatures.

For instance, dolphins never attack men. They may play games with them and pretend to bite, but they don't really. There are even three cases on record of the creatures guiding men, who have fallen overboard, back to shore. On the other hand, a dolphin, who had earlier played harmlessly with a man, promptly killed a barracuda placed in its tank with one snap of its jaws.

Now I'm not sure that it is a true sign of intelligence to play gently with men and kill barracudas, considering that man is much the more ferocious and dangerous creature. However, since man himself is acting as the judge of intelligence, there is no question but that he will give the dolphin high marks for behaving in a fashion he cannot help but approve.

Fortunately, there are more objective reasons for suspecting the existence of intelligence in dolphins. Although only five to eleven feet long, and therefore not very much more massive than men, they have a larger and more convoluted brain. It is not so much the size of the brain that counts as a measure of intelligence as the extent of its surface area, since upon that depends the quality of gray matter. As, in the course of evolution, the brain surface increased faster than space for it was made available, it had to fold into convolutions. The extent and number of convolutions increases as one goes from opossum to cat to monkey to man, but it is the cetaceans (a general term for whales, dolphins and porpoises), and *not* man, that holds the record in this respect.

Well, then, is the intelligence of the dolphin, so clearly advertised by its brain, really a new discovery? I doubt it strongly. I think the Greeks anticipated us by a few millennia.

For instance, there is the old Greek tale of Arion, a lyre-player and singer in the employ of Periander of Corinth. Arion, having won numerous prizes at a music festival in Sicily, took passage back to Corinth. While en route the honest sailors on board saw no reason why they could not earn a bonus for themselves by the simple expedient of throwing Arion overboard and appropriating his prizes.

Being men of decision and action, they were about to do exactly that when Arion asked the privilege of singing one last song. This was granted by the sailors who, after all, were killing Arion only by way of business and not out of personal animosity.

Arion's sweet song attracted a school of dolphins, and when he jumped overboard at its end, one of the dolphins took him on his back and sped him back to Corinth faster than the ship itself could travel. When the sailors arrived in port, Arion was there to bear witness against them and they were executed. The story doesn't say whether they were allowed to sing a final song before being killed.

But why a dolphin? Surely if the Greeks were merely composing a fantasy, a shark would have done as well, or a giant sea horse, or a merman, or a monstrous snail. Yet they chose a dolphin not only in the Arion tale but in several others. It seems clear to me that dolphins were chosen deliberately because Greek sailors had observed just those characteristics of the creatures that the Navy is now observing once more—their intelligence and (no other word will do) kindliness.

Oddly enough, the Greeks, or at least one Greek, was far ahead of his time in another matter which involved the dolphin.

Let me backtrack a little in order to explain. In ancient times, living creatures were classified into broad groups depending on characteristics that were as obvious as possible. For instance, anything that lived in the water permanently was a fish.

Nowadays, to be sure, we restrict the word fish to vertebrates which have scales and breathe by means of gills. Invertebrates such as clams, oysters, lobsters and crabs are *not* fish. However, the English language has never caught up with the modern subtleties of taxonomy. These non-fish are all lumped together as "shellfish." Even more primitive creatures receive the appellation, so that we speak of "starfish" and "jellyfish."

By modern definition, even a sea-dwelling vertebrate is not a fish if it lacks gills and scales, which means that whales and their smaller relatives are not fish. To the modern biologist, this seems obvious. The cetaceans are warmblooded, breathe by means of lungs and, in many ways, show clearly that they are descended from land-dwelling creatures.

However, they have become so completely adapted to the sea that they have lost any visible trace of their hind limbs, transformed their front limbs into vaguely fishlike flippers, and developed a tail that is horizontal rather than vertical but otherwise again superficially fishlike. They have even streamlined themselves into a completely fish-like shape. For all these reasons, what is obvious to the biologist is not obvious to the general public, and popular speech insists on calling a whale a fish.

Thus, the song "It Ain't Necessarily So" from *Porgy and Bess* speaks of Jonah living in a whale. Those of us who are sophisticated and know the difference between mammals and fish by the modern definition, can laugh good-humoredly at the charming simplicity of the characters in the play; but actually, the mistake is in reverse, and almost all of us are involved in it.

The book of Jonah does *not*—I repeat, does *not*—mention any whale. Jonah 1:17 (in the King James version) reads: "Now the Lord had prepared a great fish to swallow up Jonah. And Jonah was in the belly of the fish three days and three nights." The creature is mentioned again in Jonah 2:1 and Jonah 2:10, each time as a fish.

It is only folk taxonomy that converted "great fish" into "whale."

And now I am ready for the Greeks again. The first to distinguish cetaceans from other sea creatures was Aristotle, back about 340 B.C. In a book called *Generation of Animals* he pointed out something that, considering the times, was a miracle of accurate observation; that is, that dolphins brought forth young alive, and that the young dolphins, when born, were attached to their mother by an umbilical cord. Now, an umbilical cord implies that the embryo derives nourishment from the mother, continuously and directly, and not from a fixed food supply within an egg, as is the case with, for instance, certain snakes that bring forth their young alive. The umbilical cord is characteristic of the hairy, milk-yielding quadrupeds we call mammals, and of no other creatures.

For that reason, Arisotole classified the cetaceans with the mammals and not with the fish.

Frustratingly enough, where so many of Aristotle's wrong guesses and deductions were held to by ancient and medieval thinkers in a kind of death grip, this accurate observation which fits perfectly with our modern ways of thinking, was ignored. It wasn't until the nineteenth century that Aristotle's statements about the dolphin were finally confirmed.

One thing we have learned recently about porpoises, that the Ancient Greeks probably did not know, concerns the noises they make. Now microphones under the sea have shown us that the ocean (surprisingly enough) is a noisy place, with shellfish clicking their claws and fish grunting weirdly. However, the cetaceans are the only creatures, aside from man, with brains complex enough to permit the delicate muscular movements that can produce sounds in wide variety. Porpoises do, in fact, whistle and rasp and grunt and creak in all sorts of ways. What's more, they have very well developed inner ears and can hear perfectly well all the sounds they make.

Surely, the thought should occur to us at this point that if porpoises are so all-fired smart and make all those noises, then perhaps they are *talking*. After all, we might argue, what else is such a battery of sound-formation good for?

Unfortunately, there is something aside from communication that sound is good for, and this noncommunicative use definitely exists in the case of the porpoise.

To explain this, let me go a bit into the nature of sound. Here, once more, we have a case where the Greeks got started on the right track (just about the only branch of physics in which they did).

Thus, Pythagoras of Samos observed, about 500 B.C., that if two lyre strings were plucked, the shorter string emitted the higher-pitched sound. He, or some of his followers, may also have observed that both strings vibrated while sounding and that the shorter string vibrated the faster. At any rate, by 400 B.C., a philosopher of the Pythagorean school, Archytas of Tarentum, stated that sound was produced by the striking together of bodies, swift motion producing high pitch and slow motion producing low pitch.

Then Aristotle came along and specifically included air among the bodies which produced sound when struck. He stated further that one part of the air struck the next so that sound was propagated through it until it reached the ear. Without a medium such as air or water, Aristotle pointed out, man would not hear sound. And here he was correct again.

By the end of ancient times, Boethius, the last of the Roman philosophers, writing about 500 A.D., was comparing sound waves to water waves. (Actually, sound waves are longitudinal, while water waves are transverse—a distinction I won't go into further here—but the analogy is a good one in many ways.)

Long after it was accepted that sound was a wave motion, the nature of light remained in sharp dispute. During the great birth of modern science in the seventeenth century, one group, following Christian Huyghens, believed light to be a wave phenomenon, too, like sound. A larger group, however, following Isaac Newton, believed it to be a stream of very small, very fast particles.

The reason why the Newtonian corpuscular theory of light held sway for a century and more was not only because of Newton's great prestige, but because of a line of argument that can be presented most simply as follows:

Water waves (the most familiar of all wave motions) bend around obstacles. Float a stick on water in the path of an outspreading ripple and the smooth circular arc of the ripple will be disturbed and bent into a more complicated pattern, but it will not be stopped or reflected. There will be no ripple-free "shadow" cast by the stick.

Sound also is not stopped by obstacles but bends around them. After all, you can hear a friend quite distinctly if he is calling to you from around a corner or from behind a tall fence.

Light, on the other hand, does not bend around obstacles, but is reflected by them, and you can see by this reflection. Further more, the obstacle produces a light-free area behind it, a shadow with sharp edges that proves the light rays to be traveling in absolutely straight lines. This is so different from the behavior of water waves or sound waves that it seemed clear that light could not be waves but must be particles which would travel in straight lines and would not bend around them.

Then, in 1801, an English physicist, Thomas Young, allowed a narrow beam of light to pass through two closely spaced holes. The beams that resulted broadened and over-lapped on a screen behind. But instead of a simple rein-forcement of light and a consequent brightening at the region of overlap, a series of bands or fringes of light were produced, separated by bands of darkness.

Now how could two beams of light unite to give regions of darkness? These seemd no way of explaining this if light consisted of particles. However, if they consisted of waves, there would be regions where the waves of the two light beams would be in phase (that is, moving up and down together) and therefore yield a region of light brighter than either could produce separately. There would also be regions where the two waves would be out of phase (one moving up while the other moved down) so that the two would cancel and produce darkness. This phenomenon of "interference" could be duplicated in water waves to produce exactly the effect observed by Young in the case of light.

That established the wave nature of light at once. (Actually, the modern view of light has it possessing both wave and particle aspects, but we needn't bother about that here.)

Furthermore, from the width of the interference bands and the distance between the two holes through which the beams issued, it proved possible to calculate the wave length of the light waves. The shortest wave lengths (those of violet light) were as short as 0.000039 centimeters; while the longest (those of red light) attained a length of 0.000075 centimeters.

Then, in 1818, the French physicist Augustin Jean Fresnel worked out the mathematics of wave motion and showed that waves would only travel about obstructions that were small compared to the wave length. A stick does not stop a water wave, but a long spit of land will; even in a storm, a region of the sea protected from open ocean by such a spit will remain relatively calm—in a wave-free "shadow." In fact, that's the whole principle underlying the value of harbors.

Conversely, objects visible to the naked eye, however small they may be, are large compared to the wave lengths of light, and that is why light waves don't bend around them, but are reflected and cast sharp shadows instead. If the obstructions are only made small enough, light waves *will* bend around them (a phenomenon called "diffraction"); this, Fresnel was able actually to demonstrate.

Now, then, back to sound. The wave lengths of sound must be much longer than those of light, because sound is diffracted by obstacles that stop light cold. (Nevertheless, although a tree will not reflect ordinary sound waves, mountains will, thus producing echoes, and sounds in large rooms will reverberate as sound waves bounce off the walls.)

The exact wave length of a particular sound wave can be obtained by dividing the velocity of sound by the frequency (that is, the number of times the sound source vibrates per second).

As for the velocity of sound, even primitive people must have known that sound has some finite speed, because at quite moderate distances you can see a woodcutter swing his ax and hit the tree, and then only after a moment or two will you hear the "thunk." Assuming that light travels at infinite velocity (and, compared with sound, it virtually does), all you would have to do to determine the velocity of sound would be to measure the time interval between seeing and hearing over a known distance.

The difficulty of timing that interval was too much for scholars until modern times. It wasn't until 1656 that the pendulum clock was invented (by Huyghens, by the way, who originated the wave theory of light); only then did it become possible to measure intervals of less than an hour with reasonable precision.

In 1738, French scientists set up cannons on hills some

seventeen miles apart. They fired a cannon on one hill and timed the interval between flash and sound on the other; then they fired the cannon on the second hill and timed the interval on the first. (Doing it both ways and taking the average cancels the effect of the wind.) Thus, the velocity of sound was first measured. The value accepted today is 331 meters per second at 0° C., or 740 miles an hour.

The velocity of sound depends on the elasticity of air; that is, on the natural speed with which air molecules can bounce back and forth. The elasticity increases with temperature and so, therefore, does the velocity of sound, the increase coming to roughly half a meter per second for each Centigrade degree rise in temperature.

Middle C on the piano has a frequency of 264 vibrations per second; therefore its wave length is $\frac{331}{264}$ or 1.25 meters. Frequency goes up with higher pitch (as the Pythagoreans first discovered) and wave length therefore comes down. As pitch goes lower, frequency goes down and wave length up.

The lowest note on the piano has a frequency of 27.5 vibrations per second, while the highest note has a frequency of 4,224 vibrations per second. The wave length of the former is therefore $\frac{231}{27.5}$ or 12 meters, and of the latters is $\frac{331}{4,224}$ or 0.076 meters (which is equivalent to 7.6 centimeters).

Even the wide range of the piano doesn't represent the extremes of the ear's versatility. The normal human ear can hear frequencies as low as 15 vibrations per second and as high as 15,000 per second in adults and even 20,000 per second in children. This is an extreme span of over ten octaves (each octave representing a doubling of frequency) as compared with the single octave of light to which the eye is sensitive. In terms of wave length the human ear can make out a range, at the extreme, of from 22 meters down to less than 2 centimeters.

Even the highest-pitched sound we can hear, however, has a wave length about 20,000 times as long as that of red light, so that we have every reason to expect that sound

109

and light should behave quite differently with regard to obstructions.

Still, the shorter the wave length (that is, the higher pitched the sound), the more efficiently an obstacle of a particular size ought to stop and reflect a sound wave. A tree should be able to reflect a 2-centimeter sound wave, while it would have no effect at all on a 22-meter sound wave.

Why not, then, progress still further down the wavelength scale and use sounds so high-pitched that they pass the limits of audibility? (These are "ultrasonic"—"beyond sound"—frequencies.) The existence of such inaudible sound is easily demonstrated even without man-made detectors. Whistles can be made which yield ultransonic sound when they are blown. We hear nothing; but dogs, with ears capable of detecting sounds of higher frequency than ours can, come running.

The production of ultrasonic sounds in quantity first became possible as a result of a discovery made in 1880 by two brothers, Pierre and Jacques Curie. (Pierre Curie, a brilliant scientist, happened to marry a still more brilliant one—Marie, the famous Madame Curie—and is the only great scientist in history who is consistently identified as the husband of someone else.)

The Curie brothers found that quartz crystals, if cut in the proper manner, would, when slightly compressed as a result of very high pressure, develop small electric charges on their opposite faces. This is called "piezoelectricity" from a Greek word meaning "pressure." They also discovered the reverse phenomenon: if a difference in voltage is set up in metal plates held against opposite faces of the crystal, a small compression is induced in the crystal. From this it follows that if voltage is applied and removed rapidly, the crystal will expand and contract as rapidly, to produce a sound wave of that particular frequency. It will set up an ultrasonic beam if the vibration is rapid enough.

After the radio tube was developed, the production of voltages oscillating at ultrasonic frequencies became quite practical; the French physicist Paul Langevin succeeded in producing strong ultrasonic beams in 1917. World War I was going on, and he at once attempted to put to use the fact that sounds of such short wave length could be more efficiently reflected by relatively small obstacles. He used them to detect submarines under water. From the time lapse between emission of the ultrasonic pulse and detection of the echo and from the velocity of sound in water (which is over four times as great as the velocity in air, because of water's greater elasticity), the distance of the obstruction can be determined.

After World War I, this principle was put to peacetime use in detecting schools of fish and hidden icebergs, in determining the depths of the ocean and the conformation of the sea bottom, and in other ways. It went to war again in World War II, and received the name of "sonar," the abbreviation of "*so*und *na*vigation *a*nd *r*anging."

But it would seem that sonar is one field in which mankind has been anticipated by other species of creatures by many millions of years.

The bat, for instance, is a clever flier, fluttering and flitting in an erratic course. (The original meaning of the word "bat" is "to flutter rapidly," as when you "bat your eyes," and an alternative name for the creature is "flitter-mouse.") In its wobbly course, the bat catches tiny insects with precision and evades small obstructions like twigs with ease. Considering that it flies about at twilight, it is amazing that it can do so.

In 1793, the Italian scientist Lazzaro Spallanzani found that bats could catch food and avoid obstacles even in pitch darkness and even if blinded. However, they lost this ability if they were deafened.

In the early 1940's, an American physicist, G. W. Pierce, developed a device that could pick up particularly faint ultrasonic beams; and it then turned out that bats were constantly emitting not only the faint squeaks that human ears could pick up, but inaudible ultrasonic sounds with frequencies as high as 150,000 vibrations per seconds, and wave lengths, consequently, as low as 2 millimeters.

Such short wave lengths are stopped reasonably well by insects and twigs. They are reflected, and the bats pick up

111

the echoes between squeaks and guide themselves accordingly.

And this is exactly what porpoises and dolphins do also, in detecting fish rather than insects. Their larger prey makes it unnecessary for them to drive frequency so high and wave length so low. They do make use of the ultrasonic range, but accompanying that are sounds well into the audible range, too, usually described as "creaking."

Experiments at Woods Hole, Massachusetts, in 1955, showed that porpoises could pick up food fragments as small as 15 centimeters (6 inches) wide, even under conditions of complete darkness, provided they were creaking as they swam. If they were not creaking, they did not spot the fish. (One of the reasons the Navy is interested in these creatures, by the way, is that they hope to improve sonar technique by studying porpoise technique.)

This, then, is the noncommunicative use of sound referred to earlier in the article. We can speculate that perhaps the reason why sea life is so noisy is just because of the necessity of finding food and avoiding enemies in an environment where light is so limited and the sense of sight consequently so much less useful than it is on land.

But now let's ask a second question. Even if we grant that sound was first developed for purposes of sonar, a relatively simple scheme of sound-production should suffice (as in the bats). When sound-production is as elaborate as in porpoises, isn't it fair to consider it conceivable that a secondary use involving the elaboration may have developed?

To feel our way toward an answer to that question, let's consider man and certain experiments at Cornell in the early 1940's, which used blind people as well as normally sighted people who had been blindfolded.

These were made to walk down a long hall toward a fiber-board screen which might be in any position along the hall or might not be there at all, and they were to stop as soon as they were convinced the screen was just before them.

They did very well, spotting the screen almost every time some seven feet before they got to it. Most of the subjects were quite emphatic that they could somehow "feel" the approaching partition against their faces. When their heads were veiled so that the drapery would absorb any waves in the air that might be applying pressure to delicate hairs on the face, this was not found to interfere particularly with the ability to sense the screen.

However, when the ears of the subjects were efficiently plugged, the ability was lost at once. Apparently, the small echoes of footsteps or of other incidental noises gave the barrier away and, without knowing it, the men, both blind and blindfolded, were making use of the echo location principle.

The fact that human beings make use of sonar and that this, perhaps, represents the original use of the sounds we make, has not prevented us from developing sound communication secondarily to the point where it is now the prime function of our vocal cords. It is not inconceivable, then, that porpoises, with as good a brain as our own and as good ears and as good sound-making equipment (or perhaps better in each case), might not also have developed speech.

Frankly, I wish most earnestly that they have. There are a few problems mankind has that I think might be solved if we could only talk them over with some creatures who could approach matters with a fresh and objective viewpoint.

9 The Ultimate Split of the Second

Occasionally, I get an idea for something new in science; not necessarily something important, of course, but new anyway. One of these ideas is what I will devote this chapter to.

The notion came to me some time ago, when the news broke that a subatomic particle called "xi-zero" (with "xi" pronounced "ksee," if you speak Greek, and "zigh" if you speak English) had been detected for the first time. Like other particles of its general nature, it is strangely stable, having a half-life of fully a ten-billionth (10^{-10}) of a second or so.

The last sentence may seem misprinted—you may think that I meant to write "unstable"—but no! A ten-billionth of a second can be a long time; it all depends on the scale of reference. Compared to a sextillionth (10^{-23}) of a second, a ten-billionth (10^{-10}) of a second is an aeon. The difference between those two intervals of time is as that between one day and thirty billion years.

You may grant this and yet feel an onset of dizziness. The world of split-seconds and split-split-split-seconds is a difficult one to visualize. It is easy to say "a sextillionth of a second" and just as easy to say "a ten-billionth of a second"; but no matter how easily we juggle the symbols representing such time intervals, it is impossible (or *seems* impossible) to visualize either.

My idea is intended to make split-seconds more visualizable, and I got it from the device used in a realm of measurement that is also grotesque and also outside the range of all common experience—that of astronomical distances.

There is nothing strange in saying, "Vega is a very nearby star. It's not very much more than a hundred fifty trillion (.5 × 10⁻⁴) miles away."

Most of us who read science fiction are well-used to the thought that a hundred fifty trillion miles is a very small distance on the cosmic scale. The bulk of the stars in our galaxy are something like two hundred quadrillion (2×10^{17}) miles away, and the nearest full-sized outside galaxy is more than ten quintillion (10^{19}) miles away.

Trillion, quadrillion and quintillion are all legitimate number-words, and there's no difficulty telling which is larger and by how much, if you simply want to manipulate symbols. Visualizing what they mean, however, is another thing.

So the trick is to make use of the speed of light to bring the numbers down to vest-pocket size. It doesn't change the actual distance any, but it's easier to make some sort of mental adjustment to the matter if all the zeroes of the "-illions" aren't getting in the way.

The velocity of light in a vacuum is 186,274 miles per second or, in the metric system, 299,779 kilometers per second.

A "light-second," then, can be defined as that distance which light (in a vacuum) will travel in a second of time, and is equal to 186,274 miles or to 299,779 kilometers.

It is easy to build longer units in this system. A "light-minute" is equal to 60 light-seconds; a "light-hour" is equal to 60 light-minutes, and so forth, till you reach the very familiar "light-year," which is the distance which light (in a vacuum) will travel in a year. This distance is equal to 5,890,000,000,000 miles or 9,460,000,000,000 kilometers. If you are content with round numbers, you can consider a light-year equal to six trillion (6×10^{12}) miles or nine and one-half trillion (9.5×10^{12}) kilometers.

You can go on, if you please, to "light-centuries" and "light-millennia," but hardly anyone ever does. The light-year is the unit of preference for astronomic distances. (There is also the "parsec," which is equal to 3.26 light years, or roughly twenty trillion miles, but that is a unit based on a different principle, and we need not worry about it here.)

Using light-years as the unit, we can say that Vega is 27 light-years from us, and that is a small distance considering that the bulk of the stars of our galaxy are 35,000 light-years away and that the nearest full-sized outside galaxy is 2,100,000 light-years away. The difference between 27 and 35,000 and 2,100,000, given our range of experience, is easier to visualize than that between a hundred fifty trillion and two hundred quadrillion and ten quintillion, though the ratios in both cases are the same.

Furthermore, the use of the speed of light in defining units of distance has the virtue of simplifying certain connections between time and distance.

For instance, suppose an expedition on Ganymede is, at a certain time, 500,000,000 miles from earth. (The distance, naturally, varies with time, as both worlds move about in their orbits.) This distance can also be expressed as 44.8 light-minutes.

What is the advantage of the latter expression? For one thing, 44.8 is an easier number to say and handle than 500,000,000. For another, suppose our expedition is in radio communication with earth. A message sent from Ganymede to earth (or vice versa) will take 44.8 minutes to arrive. The use of light-units expresses distance *and* speed of communication at the same time.

(In fact, in a world in which interplanetary travel is a taken-for-granted fact, I wonder if the astronauts won't start measuring distance in "radio-minutes" rather than light-minutes. Same thing, of course, but more to the point.)

Then when and if interstellar travel comes to pass, making necessary the use of velocities at near the speed of light, another advantage will come about. If time dilatation exists and the experience of time is slowed at high velocities, a trip to Vega may seem to endure for only a month or for only a week. To the stay-at-homes on earth, however, who are experiencing "objective time" (the kind of time that is experienced at low velocities—strictly speaking, at zero velocity), the trip to Vega, 27 light-years distant, cannot take place in less than 27 years. A round-tripper, no matter how quickly the journey has seemed to pass for him, will find his friends on earth a minimum of 54 years older. In the same way, a trip to the galaxy of Andromeda cannot take less than 2,100,000 years of objective time, Andromeda being 2,100,000 light-years distant. Once again, time and distance are simultaneously expressed.

My idea, then, is to apply this same principle to the realm of ultra-short intervals of time.

Instead of concentrating on the tremendously long distances light can cover in ordinary units of time, why not concentrate on the tremendously short times required for light to cover ordinary units of distance?

If we're going to speak of a light-second as equal to the distance covered by light (in a vacuum) in one second and set it equal to 186,273 miles, why not speak of a "light-mile" as equal to the time required for light (in a vacuum) to cover a distance of one mile, and set that equal to $\frac{1}{186,273}$ seconds?

Why not, indeed? The only drawback is that 186,273 is such an uneven number. However, by a curious coincidence undreamed of by the inventors of the metric system, the speed of light is very close to 300,000 kilometers per second, so that a "light-kilometer" is equal to $\frac{1}{300,000}$ of a second. It comes out even rounder if you note that 3⅓ light-kilometers are equal to just about 0.00001 or 10^{-5} seconds.

Further, to get to still smaller units of time, it is only necessary to consider light as covering smaller and smaller distances.

Thus, one kilometer (10^5 centimeters) is equal to a million millimeters, and one millimeter (10^{-1} centimeters) is equal to a million millimicrons. To go one step further down, we can say that one millimicron (10^{-7} centimeters) is equal to a million fermis. (The name "fermi" has been suggested, but has not yet been officially adopted, as far as I know, for a unit of length equal to a millionth of a millimicron, or to 10^{-13} centimeters. It is derived, of course, from the late Enrico Fermi, and I will accept the name for the purposes of this chapter.)

So we can set up a little table of light-units for ultra-

117

short intervals of time, beginning with a light kilometer,* which is itself equal to only $\frac{1}{300,000}$ of a second.

1 light-kilometer	=1,000,000 light-millimeters
1 light-millimeter	=1,000,000 light-millimicrons
1 light-millimicron	=1,000,000 light-fermis

To relate these units to conventional units of time, we need only set up another short table:

3⅓ light-kilometers	10^{-5} seconds (i.e. a hundred-thousandth of a second)
3⅓ light-millimeters	10^{-11} seconds (i.e. a hundred-billionth of a second)
3⅓ light-millimicrons	10^{-17} seconds (i.e. a hundred-quadrillionth of a second)
3⅓ light-fermis	10^{-23} seconds (i.e. a hundred-sextillionth of a second)

But why stop at the light-fermi? We can proceed on downward, dividing by a million indefinitely.

Consider the fermi again. It is equal to 10^{-13} centimeters, a ten-trillionth of a centimeter. What is interesting about this particular figure, and why the name of an atomic physicist should have been suggested for the unit, is that 10^{-13} centimeters is also the approximate diameter of the various subatomic particles.

A light-fermi, therefore, is the time required for a ray of light to travel from one end of a proton to the other. The light-fermi is the time required for the fastest known motion to cover the smallest tangible distance. Until the day comes that we discover something faster than the speed of light or something smaller than subatomic particles, we are not likely ever to have to deal with an interval of time smaller than the light-fermi. As of now, the light-fermi is the ultimate split of the second.

Of course, you may wonder what can happen in the space of a light-fermi. And if something did happen in

*If it will help, one mile equals 1⅗ kilometers, and one inch equals 25½ millimeters.

that unimaginably small interval, how could we tell it didn't take place in a light-millimicron, which is also unimaginably small for all it is equal to a million light-fermis?

Well, consider high-energy particles. These (if the energy is high enough) travel with almost the speed of light. And when one of these particles approaches another at such a speed, a reaction often takes place between them, as a result of mutual "nuclear forces" coming into play.

Nuclear forces, however, are very short-range. Their strength falls off with distance so rapidly that the forces are only appreciable within one or two fermis distance of any given particle.

We have here, then, the case of two particles passing at the speed of light and able to interact only while within a couple of fermis of each other. It would only take them a couple of light-fermis to enter and leave that tiny zone of interaction at the tremendous speed at which they are moving. Yet reactions *do* take place!

Nuclear reactions taking place in light-fermis of time are classed as "strong interactions." They are the results of forces that can make themselves felt in the most evanescent imaginable interval, and these are the strongest forces we know of. Nuclear forces of this sort are, in fact, about 135 times as strong as the electromagnetic forces with which we are familiar.

Scientists adjusted themselves to this fact and were prepared to have any nuclear reactions involving individual subatomic particles take only light-fermis of time to transpire.

But then complications arose. When particles were slammed together with sufficient energy to undergo strong interactions, new particles not previously observed were created in the process and were detected. Some of these new particles (first observed in 1950) amazed scientists by proving to be very massive. They were distinctly more massive, in fact, than neutrons or protons, which, until then, had been the most massive particles known.

These super-massive particles are called "hyperons" (the prefix "hyper-" comes from the Greek and means "over," "above," "beyond"). There are three classes of these hyperons, distinguished by being given the names of different Greek letters. There are the lambda particles, which are about 12 per cent heavier than the proton; the sigma particles, which are about 13 per cent heavier; and the xi particles, which are about 14 per cent heavier.

There were theoretical reasons for suspecting that there were one pair of lambda particles, three pairs of sigma particles, and two pairs of xi particles. These differ among themselves in the nature of their electric charge and in the fact that one of each pair is an "antiparticle." One by one, each of the hyperons was detected in bubble chamber experiments; the xi-zero particle, detected early in 1959, was the last of them. The roster of hyperons was complete.

The hyperons as a whole, however, were odd little creatures. They didn't last long, only for unimaginably small fractions of a second. To scientists, however, they seemed to last very long indeed, for nuclear forces were involved in their breakdown, which should therefore have taken place in light-fermis of time.

But they didn't. Even the most unstable of all the hyperons, the sigma-zero particle, must last at least as long as a quintillionth of a second. Put that way, it sounds like a satisfactorily short period of time—not long enough to get really bored. But when the interval is converted from conventional units to light-units, we find that a quintillionth of a second is equal to 30,000 light-fermis.

Too long!

And even so, 30,000 light-fermis represent an extraordinarily short lifetime for a hyperon. The others, including the recently-discovered xi-zero particle, have half-lives of about 30,000,000,000,000 light-fermis, or 30 light-millimeters.

Since the nuclear forces bringing about the breakdown of hyperons last at least ten trillion times as long an interval of time as that required to form them, those forces must be that much weaker than those involved in the "strong interactions." Naturally, the new forces are spoken of as being involved in the "weak interactions"; and they are weak indeed, being almost a trillion times weaker than even electromagnetic forces.

In fact, the new particles which were involved in "weak

interactions" were called "strange particles," partly because of this, and the name has stuck. Every particle is now given a "strangeness number" which may be $+1$, 0, -1 or -2.

Ordinary particles such as protons and neutrons have strangeness numbers of 0; lambda and sigma particles have strangeness numbers of -1, xi particles have strangeness numbers of -2, and so forth. Exactly what the strangeness number signifies is not yet completely clear; but it can be worked with now and figured out later.

The path and the activities of the various hyperons (and of the other subatomic particles as well) are followed by their effects upon the molecules with which they collide. Such a collision usually involves merely the tearing off of an electron or two from the air molecules. What is left of the molecule is a charged "ion."

An ion is much more efficient as a center about which a water droplet can form, than is the original uncharged molecule. If a speeding particle collides with molecules in a sample of air which is supercharged with water vapor (as in a Wilson cloud chamber) or in liquid hydrogen at the point of boiling (as in a bubble chamber), each ion that is produced is immediately made the center of a water droplet or gas droplet, respectively. The moving particle marks its path, therefore, with a delicate line of water drops. When the particle breaks down into two other particles, moving off in two different directions, the line of water gives this away by splitting into a Y form.

It all happens instantaneously to our merely human senses. But photograph upon photograph of the tracks that result will allow nuclear physicists to deduce the chain of events that produce the different track patterns.

Only subatomic particles that are themselves charged are very efficient in knocking electrons out of the edges of air molecules. For that reason, only charged particles can be followed by the water traceries. And, also for that reason, in any class of particles, the uncharged or neutral varieties are the last to be detected.

For instance, the neutron, which is uncharged, was not discovered until eighteen years after the discovery of the similar, but electrically charged, proton. And in the case of the hyperons, the last to be found was xi-zero, one of the uncharged varieties. (The "zero" means "zero charge.")

Yet the uncharged particles can be detected by the very absence of a trace. For instance, the xi-zero particle was formed from a charged particle, and broke down, eventually, into another type of charged particle. In the photograph that finally landed the jackpot (about seventy thousand were examined), there were lines of droplets separated by a significant gap! That gap could not be filled by any known uncharged particle, for any of those would have brought about a different type of gap or a different sequence of events at the conclusion of the gap. Only the xi-zero could be made to fit; and so, in this thoroughly negative manner, the final particle was discovered.

And where do the light-units I'm suggesting come in? Why, consider that a particle traveling at almost the speed of light has a chance, if its lifetime is about 30 light-millimeters, to travel 30 light-millimeters before breaking down.

The one implies the other. By using conventional units, you might say that a length of water droplets of about 30 millimeters implies a half-life of about a trillionth of a second (or vice versa), but there is no obvious connection between the two numerical values. To say that a track of 30 millimeters implies a half-life of 30 light-millimeters is equally true, and how neatly it ties in. Once again, as in the case of astronomical distances, the use of the speed of light allows one number to express both distance and time.

A group of particles which entered the scene earlier than the hyperons are the "mesons." These are middleweight particles, lighter than protons or neutrons, but heavier than electrons. (Hence their name, from a Greek word meaning "middle.")

There are three known varieties of these particles, too. The two lighter varieties are also distinguished by means of Greek letters. They are the mu-mesons, discovered in 1935, which are about 0.11 as massive as a proton, and the pi-mesons, discovered in 1947, which are about 0.15 as massive as protons. Finally, beginning in 1949, various species of unusually heavy mesons, the K-mesons, were discovered. These are about 0.53 as massive as protons.

On the whole, the mesons are less unstable than the hyperons. They have longer half-lives. Whereas even the most stable of the hyperons has a half-life of only 30 light-millimeters, the meson half-lives generally range from that value up through 8,000 light-millimeters for those pi-mesons carrying an electric charge, to 800,000 light-millimeters for the mu-mesons.

By now, the figure of 800,000 light-millimeters ought to give you the impression of a long half-life indeed, so I'll just remind you that by conventional units it is the equivalent of $\frac{1}{400,000}$ of a second.

A short time to us, but a long- lo-o-o-ong time on the nuclear scale.

Of the mesons, it is only the K-variety that comes under the heading of strange particles. The K-plus and K-zero mesons have a strangeness number of $+1$, and the K-minus meson, a strangeness number of -1.

The weak interactions, by the way, recently opened the door to a revolution in physics. For the first eight years or so after their discovery, the weak interactions had seemed to be little more than confusing nuisances. Then, in 1957, as a result of studies involving them, the "law of conservation of parity" was shown not to apply to all processes in nature.

I won't go into the details of that, but it's perhaps enough to say that the demonstration thunderstruck physicists; that the two young Chinese students who turned the trick (the older one was in his middle thirties) were promptly awarded the Nobel Prize; and that a whole new horizon in nuclear theory seems to be opening up as a result.

Aside from the mesons and hyperons, there is only one unstable particle known—the neutron. Within the atomic nucleus the neutron is stable; but in isolation, it eventually breaks down to form a proton, an electron and a neutron. (Of course, antiparticles such as positrons and antiprotons are unstable in the sense that they will react with electrons and protons, respectively. Under ordinary circumstances this will happen in a millionth of a second or so. However, if these antiparticles were in isolation, they would remain as they were indefinitely, and that is what we mean by stability.)

The half-life of the neutron breakdown is 1,010 seconds (or about 17 minutes), and this is about a billion times longer than the half-life of the breakdown of any other known particle.

In light-units, the half-life of the neutron would be 350,000,000 light-kilometers. In other words, if a number of neutrons were speeding at the velocity of light, they would travel 350,000,000 kilometers (from one extreme of earth's orbit to the other, plus a little extra) before half had broken down.

Of course, neutrons as made use of by scientists don't go at anything like the speed of light. In fact, the neutrons that are particularly useful in initiating uranium fission are very slow-moving neutrons that don't move any faster than air molecules do. Their speed is roughly a mile a second.

Even at that creep, a stream of neutrons would travel a thousand miles before half had broken down. And in that thousand miles, many other things have a chance to happen to them. For instance, if they're traveling through uranium or plutonium, they have a chance to be absorbed by nuclei and to initiate fission. And to help make the confusing and dangerous—but exciting—world we live in today.

10 Order! Order!

One of the big dramatic words in science is ENTROPY. It comes so trippingly and casually to the tongue; yet when the speaker is asked to explain the term, lockjaw generally sets in. Nor do I exonerate myself in this respect. I, too, have used the word with fine abandon and have learned to change the subject deftly when asked to explain its meaning.

But I must not allow myself to be cowardly forever. So, with lips set firmly and with a face a little pale, here I go. . . .

I will begin with the law of conservation of energy. This states that energy may be converted from one form to another but can be neither created nor destroyed.

This law is an expression of the common experience of mankind. That is, no one knows any reason *why* energy cannot be created or destroyed; it is just that so far neither the most ingenious experiment nor the most careful observation has ever unearthed a case where energy is created or destroyed.

The law of conservation of energy was established in the 1840's, and it rocked along happily for half a century. It was completely adequate to meet all earthly problems that came up. To be sure, astronomers wondered whence came the giant floods of energy released by the sun throughout the long history of the solar system, and could find no answer which satisfactorily met the requirements of both astronomy and the law of conservation of energy.

However, that was the sun. There were no problems on earth—until radioactivity was discovered.

In the 1890's the problem arose of finding out from

whence came the tremendous energy released by radioactive substances. For a decade or so, the law of conservation of energy looked sick indeed. Then, in 1905, Albert Einstein demonstrated (mathematically) that mass and energy had to be different forms of the same thing, and that a very tiny bit of mass was equivalent to a great deal of energy. All the energy released by radioactivity was at the expense of the disappearance of an amount of mass too small to measure by ordinary methods. And in setting up this proposition, a source of energy was found which beautifully explained the radiation of the sun and other stars.

In the years after 1905, Einstein's theory was demonstrated experimentally over and over again, with the first atom bomb in 1945 as the grand culmination. The law of conservation of energy is now more solidly enthroned than ever, and scientists do not seriously expect to see it upset at any time in the future except under any but the most esoteric circumstances conceivable.

In fact, so solidly is the law enthroned that no patent office in its right mind would waste a split-second considering any device that purported to deliver more energy than it consumed. (This is called a "perpetual motion machine of the first class.")

The first machine that converted heat into mechanical work on a large scale was the steam engine, invented at the beginning of the eighteenth century by Thomas Newcomen and made thoroughly practical at the end of that century by James Watt. Since the steam engine produced work through the movement of energy in the form of heat from a hot reservoir of steam to a cold reservoir of water, the study of the interconversion of energy and work was named "thermodynamics" from Greek words meaning "movement of heat." The law of conservation of energy is so fundamental to devices such as the steam engine, that the law is often called the First Law of Thermodynamics.

The First Law tells us that if the reservoir of steam contains a certain amount of energy, you cannot get more work out of a steam engine than is equivalent to that much energy. That seems fair enough, perhaps; you can't get something for nothing.

But surely you can at least get out the full amount of work equivalent to the energy content of the steam, at least assuming you cut down on waste and on friction.

Alas, you can't. Though you built the most perfect steam engine possible, one without friction and without waste, you still could not convert all the energy to work. In thermodynamics, you not only can't win, you can't even break even.

The first to point this out unequivocally was a French physicist named Sadi Carnot, in 1824. He stated that the fraction of the heat energy that could be converted to work even under ideal conditions depended upon the temperature difference between the hot reservoir and the cold reservoir. The equation he presented is this.

$$\text{ideal efficiency} = \frac{T_2 - T_1}{T_2}$$

where T_2 is the temperature of the hot reservoir and T_1 the temperature of the cold. For the equation to make sense the temperatures should be given on the absolute scale. (The absolute scale of temperature is discussed in Chapter 12.)

If the hot reservoir is at a temperature of 400° absolute (127° C.) and the cold reservoir is at a temperature of 300° absolute (27° C.), then ideal efficiency is:

$$\frac{400 - 300}{400}$$

or exactly 0.25. In other words, one quarter of the heat content of the steam, *at best,* can be converted into work, while the other three-quarters is simply unused.

Furthermore, if you had only a hot reservoir and nothing else, so that it had to serve as hot and cold reservoir both, Carnot's equation would give the ideal efficiency as:

$$\frac{400 - 400}{400}$$

which is exactly zero. The steam has plenty of energy in it, but none of that energy, none at all, can be converted to work unless somewhere in the device there is a temperature difference.

An analogous situation exists in other forms of energy, and the situation may be easier to understand in cases more mundane than the heat engine. A large rock at the edge of a cliff can do work, provided it moves from its position of high gravitational potential to one of low gravitational potential, say at the bottom of the cliff. The smaller the difference in gravitational potentials (the lower the cliff), the less work the rock can be made to do in falling. If there is no cliff at all, but simply a plateau of indefinite extent, the rock cannot fall, and can do no work, even though the plateau may be six miles high.

We can say then: No device can deliver work out of a system at a single energy potential level.

This is one way of stating what is called the Second Law of Thermodynamics.

A device that purports to get work out of a single energy potential level is a "perpetual motion machine of the second class." Actually almost all perpetual motion machines perpetrated by the army of misguided gadgeteers that inhabit the earth are of this type, and patent offices will waste no time on these either.

Given energy at two different potential levels, it is the common experience of mankind to observe that the energy will flow from one potential (which we can call the higher) to another (which we can call the lower) and never vice versa (unless it is pushed). In other words, heat will pass spontaneously from a hot body to a cold one; a boulder will fall spontaneously from a cliff top to a cliff bottom; an electric current will flow spontaneously from cathode to anode.

To say: "Energy will always flow from a high potential level to a low potential level" is another way of stating the Second Law of Thermodynamics. (It can be shown that an energy flow from high to low implies the fact that you cannot get work out of a one-potential system, and vice versa, so both are equivalent ways of stating the Second Law.)

Now, work is never done instantaneously. It invariably occupies time. What happens during that time? Let's suppose, for simplicity's sake, that a steam engine is functioning as a "closed system"; that is, as a sort of walled-off portion of the universe into which no energy can enter and from which no energy can depart. In such a closed-system steam engine, the Second Law states that heat must be flowing from the point of high energy potential (the hot reservoir in this case) to the point of low (the cold reservoir).

As this happens, the hot reservoir cools and the cold reservoir warms. The temperature difference between hot and cold therefore decreases during the time interval over which work is being extracted. But this means that the amount of energy which can be converted to work (an amount which depends on the size of the temperature difference) must also decrease.

Conversely, the amount of energy which *cannot* be converted into work must increase. This increase in the amount of unavailable energy is the inevitable consequence of the heat flow predicted by the Second Law. Therefore to say that in any spontaneous process (one in which energy flows from high to low) the amount of unavailable energy increases with time, is just another way of stating the Second Law.

A German physicist, Rudolf Clausius, pointed all this out in 1865. He invented a quantity consisting of the ratio of heat change to temperature and called that quantity "entropy." Why he named it that is uncertain. It comes from Greek words meaning "a change in" but that seems insufficient.

In every process involving energy change, Clausius' "entropy" increases. Even if the energy levels don't approach each other with time (the cliff top and cliff bottom don't approach each other appreciably while a rock is falling), there is always some sort of resistance to the change from one energy potential to another. The falling body must overcome the internal friction of the air it falls through, the flowing electric current must overcome the resistance of the wire it flows through. In every case, the amount of energy available for conversion to work decreases and

129

the amount of energy unavailable for work increases. And in every case this is reflected in an increase in the heat change to temperature ratio.

To be sure, we can imagine ideal cases where this doesn't happen. A hot and cold reservoir may be separated by a perfect insulator, a rock may fall through a perfect vacuum, an electric current may flow through a perfect conductor, all surfaces may be perfectly frictionless, perfectly nonradiating. In all such cases there is no entropy increase; the entropy change is zero. However such cases generally exist only in imagination; in real life the zero change in entropy may be approached but not realized. And, of course, even in the ideal case, entropy change is never negative. There is never a *decrease* in entropy.

With all this in mind, the briefest way I know to state the First and Second Laws of Thermodynamics is this:

In any closed system the total energy content remains constant while the total entropy continually increases with time.

The main line of development of the First and Second Laws took place through a consideration of heat flow alone, without regard for the structure of matter. However, the atomic theory had been announced by John Dalton in 1803, and by the mid-nineteenth century it was well-enough established for a subsidiary line of development to arise in which energy changes were interpreted by way of the movement of atoms and molecules. This introduced a statistical interpretation of the Second Law which threw a clearer light on the question of entropy decrease.

Clausius himself had worked out some of the consequences of supposing gases to be made up of randomly moving molecules, but the mathematics of such a system was brought to a high pitch of excellence by the Scottish mathematician Clerk Maxwell and the Austrian physicist Ludwig Boltzmann in 1859 and the years immediately following.

As a result of the Maxwell-Boltzmann mathematics, gases (and matter generally) could be viewed as being made up of molecules possessing a range of energies. This energy is expressed in gases as random motion, with consequent molecular collisions and rebounds, the rebounds being assumed as involving perfect elasticity so that no energy is lost in the collisions.

A particular molecule in a particular volume of gas might have, at some particular time, any amount of energy of motion ("kinetic energy") from very small amounts to very high amounts. However, over the entire volume of gas there would be some average kinetic energy of the constituent molecules, and it is this average kinetic energy which we measure as temperature.

The average kinetic energy of the molecules of a gas at 500° absolute is twice that of the molecules of a gas at 250° absolute. The kinetic energy of a particular molecule in the hot gas may be lower at a given time than that of a particular molecule in the cold gas, but the average is in direct proportion to the temperature at all times. (This is analogous to the situation in which, although the standard of living in the United States is higher than that of Egypt, a particular American may be poorer than a particular Egyptian.)

Now suppose that a sample of hot gas is brought into contact with a sample of cold gas. The average kinetic energy of the molecules of the hot gas being higher than that of the molecules of the cold gas, we can safely suppose that in the typical case, a "hot" molecule will be moving more quickly than a "cold" molecule. (I must put "hot" and "cold" in quotation marks, because such concepts as heat and temperature do not apply to individual molecules but only to systems containing large numbers of molecules.)

When the two rebound, the total energy of the two does not change, but it may be distributed in a new way. Some varieties of redistribution will involve the "hot" molecule growing "hotter" while the "cold" molecule grows "colder" so that, in effect, the high-energy molecule gains further energy at the expense of the low-energy particle. There are many, many more ways of redistribution, however, in which the low-energy particle gains energy at the expense of the high-energy particle, so that both end up with intermediate energies. The "hot" molecule grows "cooler" while the "cold" molecule grows "warmer."

This means that if a very large number of collisions are considered, the vast majority of those collisions will result in a more even distribution of energy. The few cases in which the energy difference becomes more extreme will be completely swamped. On the whole, the hot gas will cool, the cold gas will warm, and eventually the two gases will reach equilibrium at some single (and intermediate) temperature.

Of course, it *is* possible, in this statistical view, that through some quirk of coincidence, all the "hot" molecules (or almost all) will just happen to gain energy from all the "cold" molecules, so that heat will flow from the cold body to the hot body. This will increase the temperature difference and, consequently, the amount of energy available for conversion into work, and thus will *decrease* the unavailable energy, which we call entropy.

Statistically speaking, then, a kettle of water *might* freeze while the fire under it grows hotter. The chances of this (if worked out from Maxwell's mathematics) are so small, however, that if the entire known universe consisted of nothing but kettles of water over fires, then the probability that any single one of them would freeze during the entire known duration of the universe would be so small that we could not reasonably hope to witness even a single occurrence of this odd phenomenon if we had been watching every kettle every moment of time.

For that matter, the molecules of air in an empty stoppered bottle are moving in all directions randomly. It *is* possible that, by sheer chance, all molecules might happen to move upward simultaneously. The total kinetic energy would then be more than sufficient to overcome gravitational attraction and the bottle would spontaneously fall upward. Again the chances for this are so small that no one expects ever to see such a phenomenon.

Still it must be said that a truer way of putting the Second Law would be: "In any closed system, entropy invariably increases with time—or, at least, almost invariably."

It is also possible to view entropy as having something to do with "order" and "disorder." These words are hard to define in any foolproof way; but intuitively, we picture "order" as characteristic of a collection of objects that is methodically arranged according to a logical system. Where no such logical arrangement exists, the collection of objects is in "disorder."

Another way of looking at it is to say that something which is "in order" is so arranged that one part can be distinguished from another part. The less clear the distinction, the less "orderly" it is, and the more "disorderly."

A deck of cards which is arranged in suits, and according to value within each suit, is in perfect order. Any part of the deck can be distinguished from any other point. Point to the two of hearts and I will know it is the fifteenth card in the deck.

If the deck of cards is arranged in suits, but not according to value within the suits, I am less well off. I know the two of hearts is somewhere between the fourteenth and twenty-sixth cards, but not exactly where within that range. The distinction between one part of a deck and another has become fuzzier, so the deck is now less orderly.

If the cards are shuffled until there is no system that can be devised to predict which card is where, then I can tell nothing at all about the position of the two of hearts. One part of the deck cannot be distinguished from any other part and the deck is in complete disorder.

Another example of order is any array of objects in some kind of rank and file, whether it be atoms or molecules within a crystal, or soldiers marching smartly past a reviewing stand.

Suppose you were watching marching soldiers from a reviewing stand. If they were marching with perfect precision, you would see a row of soldiers pass, then a blank space, then another row, then another blank space, and so on. You could distinguish between two kinds of volumes of space within the marching columns, soldier-full and soldier-empty, in alternation.

If the soldiers fell somewhat out of step so that the lines grew ragged, the hitherto empty volumes would start containing a bit of soldier, while the soldier-filled volumes would be less soldier-filled. There would be less distinction possible between the two types of volumes, and the situation would be less orderly. If the soldiers were completely out of step, each walking forward at his own rate, all passing volumes would tend to be equally soldier-full, and the distinction would be even less and the disorder consequently more.

The disorder would not yet be complete, however. The soldiers would be marching in one particular direction, so if you could not distinguish one part of the line from another, you could at least still distinguish your own position with respect to them. From one position you would see them march to your left, from another to your right, from still another they would be marching toward you, and so on.

But if soldiers moved randomly, at any speed and in any direction; then, no matter what your position, some would be moving toward you, some away, some to your left, some to your right, and in all directions in between. You could make no distinctions among the soldiers or among your own possible positions, and disorder would be that much more nearly complete.

Now let's go back to molecules, and consider a quantity of hot gas in contact with a quantity of cold gas. If you could see the molecules of each you could distinguish the first from the second by the fact that the molecules of the first are moving, on the average, more quickly than those of the second. Without seeing the molecules, you can achieve a similar distinction by watching the mercury thread of a thermometer.

As heat flows from the hot gas to the cold gas, the difference in average molecular motion, and hence in temperature, decreases, and the distinction between the two becomes fuzzier. Finally, when the two are at the same temperature, no distinction is possible. In other words, as heat has flowed in the direction made necessary by the Second Law, disorder has increased. Since entropy has also increased, we might wonder if entropy and disorder weren't very analogous concepts.

Apparently, they are. In any spontaneous operation, entropy increases and so does disorder. If you shuffle an ordered deck of cards you get a disordered deck, but not vice versa. (To be sure, by a fiendish stroke of luck you might begin with a disordered deck and shuffle it into a perfect arrangement, but would you care to try it and see how long that takes you? And that task involves only fifty-two objects. Imagine the same thing with several quintillion quintillion and you will not be surprised that a kettle of water over a fire never grows cooler.)

Again, if soldiers in rank and file are told to break ranks, they will quickly become a disorderly mass of humans. It is extremely unlikely, on the other hand, that a disorderly mass of humans should, by sheer luck, suddenly find themselves marching in perfect rank and file.

No, indeed. As nearly as can be told, all spontaneous processes involve an increase in disorder and an increase in entropy, the two being analogous.

It can be shown that of all forms of energy, heat is the most disorderly. Consequently, in all spontaneous processes involving types of energy other than heat, some non-heat energy is always converted to heat, this in itself involving an increase in disorder and hence in entropy.

Under no actual conditions, however, can all the heat in a system be converted to some form of non-heat energy, since that in itself would imply an increase in order and hence a decrease in entropy. Instead, if some of the heat undergoes an entropy decrease and is converted to another form of energy, the remaining heat must undergo an entropy increase that more than makes up for the first change. The net entropy change over the *entire* system is an increase.

It is, of course, easy to cite cases of entropy decrease as long as we consider only parts of a system and not all of it. For instance, we see mankind extract metals from ores and devise engines of fiendish complexity out of metal ingots. We see elevators moving upwards and automobiles driving uphill, and soldiers getting into marching formation and cards placed in order. All these and a very large number of other operations involve decreases in entropy brought about by the action of life. Consequently, the feeling arises that life can "reverse entropy."

However, there is more to consider. Human beings are eating food and supporting themselves on the energy gained from chemical changes within the body. They are burning coal and oil to power machinery. They use hydro-electric power to form aluminum. In short, all the entropy-decreasing activities of man are at the expense of the entropy-increasing activities involved in food and fuel consumption, and the entropy-increasing activities far outweigh the entropy-decreasing activities. The net change is an entropy increase.

No matter how we might bang our gavels and cry out "Order! Order!" there is no order, there is only increasing disorder.

In fact, in considering entropy changes on earth, it is unfair to consider the earth alone, since our planet is gaining energy constantly from the sun. This influx of energy from the sun powers all those processes on earth that represent local decreases in entropy: the formation of coal and oil from plant life, the circulation of atmosphere and ocean, the raising of water in the form of vapor, and so on. It is for that reason we can continue to extract energy by burning oil and coal, by making use of power from wind, river currents, waterfalls, and so forth. It is all at the expense, indirectly, of the sun.

The entropy increase represented by the sun's large-scale conversion of mass to energy simply swamps the comparatively tiny entropy decreases on earth. The net entropy change of the solar system as a whole is that of a continuing huge increase with time.

Since this must be true of all the stars, nineteenth-century physicists reasoned that entropy in the universe as a whole must be increasing rapidly and that the time must come when the finite supply of energy in a finite universe must reach a state of maximum entropy.

In such a condition, the universe would no longer contain energy capable of being converted into useful work. It would be in a state of maximum disorder. It would be a homogeneous mass containing no temperature differences. There would be no changes by which to measure time, and therefore time would not exist. There would be no way of distinguishing one point in space from another and so space would not exist. Such an entropy maximum has been referred to as the "heat-death" of the universe.

But, of course, this presupposes that the universe is finite. If it were infinite, the supply of energy would be infinite and it would take all eternity for entropy to reach a maximum. Besides that, how can we be certain that the laws of thermodynamics worked out over small volumes of space in our laboratories and observed to be true (or to seem true) in the slightly larger volumes of our astronomic neighborhood, are true where the universe as a whole is concerned?

Perhaps there are processes we know nothing about as yet, which decrease entropy as quickly as it is increased by stellar activity, so that the net entropy change in the universe as a whole is zero. This might still be so even if we allow that small portions of space, such as single galaxies, might undergo continuous entropy increases and might eventually be involved in a kind of local heat-death. The theory of continuous creation (see "Here It Comes; There It Goes," in *Fact and Fancy*, Doubleday, 1962) does, in fact, presuppose a constant entropy level in the universe as a whole.

And even if the universe were finite, and even if it were to reach "heat-death," would that really be the end?

Once we have shuffled a deck of cards into complete randomness, there will come an *inevitable* time, *if we wait long enough*, when continued shuffling will restore at least a partial order.

Well, waiting "long enough" is no problem in a universe at heat-death, since time no longer exists then. We can therefore be certain that after a timeless interval, the purely random motion of particles and the purely random flow of energy in a universe at maximum entropy might, here and there, now and then, result in a partial restoration of order.

It is tempting to wonder if our present universe, large as it is and complex though it seems, might not be merely the result of a very slight random increase in order over a very small portion of an unbelievably colossal universe which is virtually entirely in heat-death.

Perhaps we are merely sliding down a gentle ripple that has been set up, accidentally and very temporarily, in a quiet pond, and it is only the limitation of our own infinitesimal range of viewpoint in space and time that makes it seem to ourselves that we are hurtling down a cosmic waterfall of increasing entropy, a waterfall of colossal size and duration.

11 The Modern Demonology

You would think, considering my background, that had I ever so slight a chance to drag fantasy into any serious discussion of science, I would at once do so by neon lights flashing and fireworks blasting.

And yet, in the previous chapter on entropy, I completely ignored the most famous single bit of fantasy in the history of science. Yet that was only that I might devote another entire chapter to it.

When a hot body comes into contact with a cold body, heat flows spontaneously from the hot one to the cold one and the two bodies finally come to temperature equilibrium at some intermediate level. This is one aspect of the inevitable increase of entropy in all spontaneous processes involving a closed system.

In the early nineteenth century, the popular view was to consider heat a fluid that moved from hot to cold as a stone would fall from high to low. Once a stone was at the valley bottom, it moved no more. In the same way, once the two bodies reached temperature equilibrium, there could be no further heat flow under any circumstances.

In the mid-nineteenth century, however, the Scottish mathematician James Clerk Maxwell adopted the view that temperature was the measure of the average kinetic energy of the particles of a system. The particles of a hot body moved (on the average) more rapidly than did the particles of a cold body. When such bodies were in contact, the energies were redistributed. On the whole, the most probable redistribution was for the fast particles to lose velocity (and, therefore, kinetic energy) and the slow particles to gain it. In the end, the average velocity would

be the same in both bodies and would be at some intermediate level.

In the case of this particle-in-motion theory, it *was*
conceivable for heat flow to continue after equilibrium had
been reached.

Imagine, for instance, two containers of gas connected
by a narrow passage. The entire system is at a temperature
equilibrium. That is, the average energy of the molecules
in any one sizable portion of it (a portion large enough
to be visible in an ordinary microscope) is the same as
that in any other sizable portion.

This doesn't mean that the energies of all individual
molecules are equal. There are some fast ones, some very
fast ones, some very very fast ones. There are also some
slow ones, some very slow ones and some very very slow
ones. However, they all move about higgledy-piggledy and
keep themselves well scrambled. Moreover, they are also
colliding among themselves millions of times a second so
that the velocities and energies of any one molecule are
constantly changing. Therefore, any sizable portion of the
gas has its fair share of both fast and slow molecules and
ends with the same temperature as any other sizable portion.

However, what if—just as a matter of chance—a number of high-energy molecules happened to move through
the connecting passageway from right to left while a number of low-energy molecules happened to move through
from left to right? The left container would then grow
hot and the right container cold (though the average temperature overall would remain the same). A heat-flow
would be set up despite equilibrium, and entropy would
decrease.

Now, there is a certain infinitesimal chance, unimaginably close to zero, that this would happen through the
mere random motion of molecules. The difference between
"zero" and "almost-almost-almost-zero" is negligible in
practice, but tremendous from the standpoint of theory;
for the chance of heat-flow at equilibrium is zero in the
fluid theory and almost-almost-almost-zero in the particle-
in-motion theory.

Maxwell had to find some dramatic way to emphasize this difference to the general public.

Imagine, said Maxwell, that a tiny demon sat near the passage connecting the two containers of gas. Suppose he let fast molecules pass through from right to left but not vice versa. And suppose he let slow molecules through from left to right but not vice versa. In this way, fast molecules would accumulate in the left and slow ones in the right. The left half would grow hot and the right cold. Entropy would be reversed.

The demon, however, would be helpless if heat were a continuous fluid—and in this way Maxwell successfully dramatized the difference in theories.

Maxwell's demon also dramatized the possibility of escaping from the dreadful inevitability of entropy increase. As I explained in the previous chapter, increasing entropy implies increasing disorder, a running down, a using up.

If entropy must constantly and continuously increase, then the universe is remorselessly running down, thus setting a limit (a long one, to be sure) on the existence of humanity. To some human beings, this ultimate end poses itself almost as a threat to their personal immortality, or as a denial of the omnipotence of God. There is, therefore, a strong emotional urge to deny that entropy *must* increase.

And in Maxwell's demon, they find substance for their denial. To be sure, the demon does not exist, but his essential attribute is his capacity to pick and choose among the moving molecules. Mankind's scientific ability is constantly increasing, and the day may come when he will be able, by some device, to duplicate the demon's function. Would he then not be able to decrease entropy?

Alas, there is a flaw in the argument. I hate to say this, but Maxwell cheated. The gas cannot be treated as an isolated system in the presence of the demon. The whole system would then consist of the gas *plus the demon*. In the process of selecting between fast and slow molecules, the demon's entropy would have to increase by an amount

141

that more than made up for the decrease in entropy that he brings about in the gas.

Of course, I know that you suspect I have never really studied demons of any type, let alone one of the Maxwell variety. Nevertheless, I am confident of the truth of my statement, for the whole structure of scientific knowledge requires that the demon's entropy behave in this fashion.

And if man ever invents a device that will duplicate the activity of the demon, then you can bet that that device will undergo an entropy increase greater than the entropy decrease it will bring about. You will be perfectly safe to grant any odds at all.

The cold fact is that entropy increase cannot be beaten. No one has ever measured or demonstrated an overall entropy decrease anywhere in the universe under any circumstances.

But entropy is strictly applicable only to questions of energy flow. It can be defined in precise mathematical form in relation to heat and temperature and is capable of precise measurement where heat and temperature are concerned. What, then, if we depart from the field where entropy is applicable and carry the concept elsewhere? Entropy will then lose its rigorous nature and become a rather vague measure of orderliness or a rough indicator of the general nature of spontaneous change.

If we do that, can we work up an argument to demonstrate anything we can call an entropy decrease in the broad sense of the term?

Here's an example brought up by a friend of mine during an excellently heated evening of discourse. He said:

"As soon as we leave the world of energy, it is perfectly possible to decrease entropy. Men do it all the time. Here is *Webster's New International Dictionary*. It contains every word in *Hamlet* and *King Lear* in a particular order. Shakespeare took those words, placed them in a different order and created the plays. Obviously, the words in the plays represent a much higher and more significant degree of order than do the words in the dictionary. Thus they represent, in a sense, a decrease in entropy. Where is the corresponding increase in entropy in Shakespeare?

142

He ate no more, expended no more energy, than if he had spent the entire interval boozing at the Mermaid Tavern."

He had me there, I'm afraid, and I fell back upon a shrewd device I once invented as a particularly ingenious way out of such a dead end. I changed the subject.

But I returned to it in my thoughts at periodic intervals ever since. Since I feel (intuitively) that entropy increase is a universal necessity, it seemed to me I ought to be able to think up a line of argument that would make Shakespeare's creations of his plays an example of it.

And here's the way the matter now seems to me.

If we concentrate on the words themselves, then let's remember that Shakespeare's words make sense to us only because we understand English. If we knew only Polish, a passage of Shakespeare and a passage of the dictionary would be equally meaningless. Since Polish makes use of the Latin alphabet just as English does and since the letters are in the same order, it follows, however, that a Polish-speaking individual could find any English word in the dictionary without difficulty (even if he didn't know its meaning) and could find the same word in Shakespeare only by good fortune.

Therefore the words, considered only as words, are in more orderly form in the dictionary, and if the word order in Shakespeare is compared with the word order in the dictionary, the construction of the plays represents an increase in entropy.

But in concentrating on the words as literal objects (a subtle pun, by the way), I am, of course, missing the point. I do that only to remove the words themselves from the argument.

The glory of Shakespeare is not the physical form of the symbols he uses but the ideas and concepts behind those symbols. Let Shakespeare be translated into Polish and our Polish-speaking friend would far rather read Shakespeare than a Polish dictionary.

So let us forget words and pass on to ideas. If we do that, then it is foolish to compare Shakespeare to the dictionary. Shakespeare's profound grasp of the essence of humanity came not from any dictionary but from his observation and understanding of human beings.

If we are to try to detect direction of entropy change, then, let us not compare Shakespeare's words to those in the dictionary, but Shakespeare's view of life to life itself.

Granted that no one in the history of human literature has so well interpreted the thoughts and emotions of mankind as well as Shakespeare has, it does not necessarily follow that he has improved on life itself.

It is simply impossible, in any cast of characters fewer than all men who have ever existed, in any set of passions weaker or less complex and intertangled than all that have ever existed, completely to duplicate life. Shakespeare has had to epitomize, and has done that superlatively well. In a cast of twenty and in the space of three hours, he exhibits more emotion and a more sensitive portrayal of various facets of humanity than any group of twenty real people could possibly manage in the interval of three real hours. In that respect he has produced what we might call a local decrease in entropy.

But if we take the entire system, and compare all of Shakespeare to all of life, surely it must be clear that Shakespeare has inevitably missed a vast amount of the complexity and profundity of the human mass and that his plays represent an overall increase of entropy.

And what is true for Shakespeare is true for all mankind's intellectual activity, it seems to me.

How I can best put this I am not certain, but I feel that nothing the mind of man can create is truly created out of nothing. All possible mathematical relationships; natural laws; combinations of words, lines, colors, sounds; all—everything—exists at least in potentiality. A particular man discovers one or another of these but does not create them in the ultimate sense of the word.

In seizing the potentiality and putting it into the concrete, there is always the possibility that something is lost in the translation, so to speak, and that represents an entropy increase.

Perhaps very little is lost, as for instance in mathematics. The relationship expressed by the Pythagorean theorem existed before Pythagoras, mankind, and the earth. Once grasped, it was grasped as it was. I don't see what can have been significantly lost in the translation. The energy increase is virtually zero.

In the theories of the physical sciences, there is clearly less perfection and therefore a perceptible entropy increase. And in literature and the fine arts, intended to move our emotions and display us to ourselves, the entropy increase—even in the case of transcendent geniuses such as Sophocles and Beethoven—must be vast.

And certainly there is never an improvement on the potentiality; there is never a creation of that which has no potential existence. Which is a way of saying there is never a decrease in entropy.

I could almost wish, at this point, that I were in the habit of expressing myself in theological terms, for if I were, I might be able to compress my entire thesis into a sentence.

All knowledge of every variety (I might say) is in the mind of God—and the human intellect, even the best, in trying to pluck it forth can but "see through a glass, darkly."

Another example of what appears to be steadily decreasing entropy on a grand scale lies in the evolution of living organisms.

I don't mean by this the fact that organisms build up complex compounds from simple ones or that they grow and proliferate. This is done at the expense of solar energy, and it is no trick at all to show that an overall entropy increase is involved.

There is a somewhat more subtle point to be made. The specific characteristics of living cells (and therefore of living multicellular organisms, too, by way of the sex cells) are passed on from generation to generation by duplication of genes. The genes are immensely complicated compounds and, ideally, the duplication should be perfect.

But where are ideals fulfilled in this imperfect universe of ours? Errors will slip in, and these departures from perfection in duplication are called mutations. Since the errors are random and since there are many more ways in which a very complex chemical can lose complexity rather than gain it, the large majority of mutations are for the worse, in the sense that the cell or organism loses a capacity that its parent possessed.

(By analogy, there are many more ways in which a hard jar is likely to damage the workings of a delicate watch than to improve them. For that reason do not hit a stopped watch with a hammer and expect it to start again.)

This mutation-for-the-worse is in accord with the notion of increasing entropy. From generation to generation, the original gene pattern fuzzes out. There is an increase of disorder, each new organism loses something in the translation, and life degenerates to death. This should inevitably happen if only mutations are involved.

Yet this does not happen.

Not only does it not happen, but the reverse *does* happen. On the whole, living organisms have grown more complex and more specialized over the aeons. Out of unicellular creatures came muticellular ones. Out of two germ layers came three. Out of a two-chambered heart came a four-chambered one.

This form of apparent entropy-decrease cannot be explained by bringing in solar energy. To be sure, an input of energy in reasonable amounts (short of the lethal level, that is) will increase the mutation rate. But it will not change the ratio of unfavorable to favorable changes. Energy input would simply drive life into genetic chaos all the faster.

The only possible way out is to have recourse to a demon (after the fashion of Maxwell) which is capable of picking and choosing among mutations, allowing some to pass and others not.

There is such a demon in actual fact, though, as far as I know, I am the only one who has called it that and drawn the analogy with Maxwell's demon. The English naturalist Charles Robert Darwin discovered the demon, so we can call it "Darwin's demon" even though Darwin himself called it "natural selection."

Those mutations which render a creature less fit to compete with other organisms for food, for mating or for self-defense, are likely to cause that creature to come to an untimely end. Those mutations which improve the creature's competing ability are likely to cause that creature to flourish. And, to be sure, fitness or lack of it relates only to the particular environment in which the creature finds itself. The best fins in the world would do a camel no good.

The effect of mutation *in the presence of natural selection*, then, is to improve continually the adjustment of a particular creature to its particular environment; and that is the direction of increasing entropy.

This may sound like arbitrarily defining entropy increase as the opposite of what it is usually taken to be—allowing entropy increase to signify increased order rather than increased disorder. This, however, is not so. I will explain by analogy.

Suppose you had a number of small figurines of various shapes and sizes lined up in orderly rank and file in the center of a large tray. If you shake the tray, the figurines will move out of place and become steadily more disordered.

This is analogous to the process of mutation without natural selection. Entropy obviously increases.

But suppose that the bottom of the tray possessed depressions into which the various figurines would just fit. If the figurines were placed higgledy-piggledy on the tray with not one figurine within a matching depression, then shaking the tray would allow each figurine to find its own niche and settle down into it.

Once a figurine found its niche through random motion, it would take a hard shake to throw it out.

This is analogous to the process of mutation with natural selection. Here entropy increases, for each figurine would have found a position where its center of gravity is lower than it would be in any other nearby position. And lowering the center of gravity is a common method of increasing entropy as, for instance, when a stone rolls downhill.

The organisms with which we are best acquainted have improved their fit to their environment by an increase in complexity in certain particularly noticeable respects. Consequently, we commonly think of evolution as necessarily proceeding from the simple to the complex.

This is an illusion. Where a simplifying change improves the fit of an organism to its environment, there the direction of evolution is from the complex to the simple. Cave creatures who live in utter darkness usually lose their eyes, although allied species living in the open retain theirs.

The reptiles went to a lot of trouble (so to speak) to develop two pairs of legs strong enough to lift the body clear of the ground. The snakes gave up those legs, slither on abdominal scales, and are the most successful of the contemporary reptiles.

Parasites undergo particularly great simplifications. A tapeworm suits itself perfectly to its environment by giving up the digestive system it no longer needs, the locomotor functions it doesn't use. It becomes merely an absorbing surface with a hooked proboscis with which to catch hold of the intestinal lining of its host, and the capacity to produce eggs and eggs and eggs and . . .

Such changes are usually called (with more than a faint air of disapproval) "degenerative." That, however, is only our prejudice. Why should we approve of some adjustments and disapprove of others? To the cold and random of evolution, an adjustment is an adjustment.

If we sink to the biochemical level, then the human being has lost a great many synthetic abilities possessed by other species and, in particular, by plants and microorganisms. Our loss of ability to manufacture a variety of vitamins makes us dependent on our diet and, therefore, on the greater synthetic versatility of other creatures. This is as much a "degenerative" change as the tapeworm's abandonment of a stomach it no longer needs, but since we are prejudiced in our own favor, we don't mention it.

And, of course, no adjustment is final. If the environment changes; if the planetary climate becomes markedly colder, warmer, drier or damper; if a predator improves its efficiency or a new predator comes upon the scene; if a parasitic organism increases in infectivity or virulence; if the food supply dwindles for any reason—then an adjustment that was a satisfactory one before becomes an unsatisfactory one now and the species dies out.

The better the fit to a particular environment, the smaller the change in environment required to bring about extinction. Long-lived species are therefore those which pick a particularly stable environment; or those that remain somewhat generalized, being fitted well enough to one environment to compete successfully within it, but not so well as to be unable to shift to an allied environment if the first fails them.

In the case of Darwin's demon (as in that of Maxwell's demon), the question as to the role of human intelligence arises. Here it is not a matter of imitating the demon, but, rather, of stultifying it.

Many feel that the advance of human technology hampers the working of natural selection. It allows people with bad eyes to get along by means of glasses, diabetics to get along by means of insulin injections, the feeble-minded to get along by means of welfare agencies, and so on.

Some people call this "degenerative mutation pressure" and, as you can see from the very expression used, are concerned about it. Everyone without exception, as far as I know, considers this a danger to humanity, even though practically nobody proposes any non-humane solutions.

And yet is it necessarily a danger to humanity?

Let's turn degenerative mutation pressure upside down and see if it can't be viewed as something other than a danger.

In the first place, we can't really stultify Darwin's demon, for natural selection must work at all times, by definition. *Man is part of nature* and his influence is as much a natural one as is that of wind and water.

So let us assume that natural selection is working and ask what it is doing. Since it is fitting man to his environment (the only thing Darwin's demon can or does do), we must inquire as to what man's environment is. In a sense, it is all the world, from steaming rain jungle to frozen glacier, and the reason for that is that all contemporary men, however primitive, band together into societies that can more or less change the environment to suit their needs even if only by building a campfire or chipping a rock or tearing off a tree branch.

Consequently, it seems clear that the most important

part of a man's environment is other men; or, if you prefer, human society. The vast majority of mankind, in fact, live as part of very complex societies that penetrate every facet of their lives.

If nearsightedness is not the handicap in New York that it would have been in a primitive hunting society, or if diabetes is not the handicap in Moscow that it would have been in a non-biochemical society, then why should there be any evolutionary pressure in favor of keeping unnecessarily good eyes and functional pancreases?

Man is to an increasing extent a parasite on human society; perhaps what we call "degenerative mutation pressure" is simply better fitting him to his new role, just as it better fit the tapeworm to its role. We may not like it, but it is a reasonable evolutionary change.

There are many among us who chafe at the restrictions of the crowded anthills we call cities, at the slavery to the clock hand, at the pressures and tensions. Some revolt by turning to delinquency, to "antisocial behavior." Others search out the dwindling areas where man can carry on a pioneer existence.

But if our anthills are to survive, we need those who will bend to its needs, who will avoid walking on grass, beating red lights and littering sidewalks. It is just the metabolically handicapped that can be relied on to do this, for they cannot afford to fight a society on which they depend, very literally, for life. A diabetic won't long for the great outdoors if it means his insulin supply will vanish.

If this is so, then Darwin's demon is only doing what comes naturally.

But of all environments, that produced by man's complex technology is perhaps the most unstable and rickety. In its present form, our society is not two centuries old, and a few nuclear bombs will do it in.

To be sure, evolution works over long periods of time and two centuries is far from sufficient to breed Homo technikos.

The closer this is approached, however, the more dangerous would become any shaking of our social structure. The destruction of our technological society in a fit of nuclear peevishness would become disastrous even if there were many millions of immediate survivors.

The environment toward which they were fitted would be gone, and Darwin's demon would wipe them out remorselessly and without a backward glance.

12 The Height of Up

Most of us would consider the surface of the sun to be pretty hot. Its temperature, as judged by the type of radiation it emits, is about 6,000° K. (with "K." standing for Kelvin scale of temperature).

However, Homo sapiens, with his own hot little hands, can do better than that. He has put together nuclear fission bombs which can easily reach temperatures well beyond 100,000° K.

To be sure, though, nature isn't through. The sun's corona has an estimated temperature of about 1,000,000° K., and the center of the sun is estimated to have a temperature of about 20,000,000° K.

Ah, but man can top that, too. The hydrogen bomb develops temperatures of about 100,000,000° K.

And yet nature still beats us, since it is estimated that the central regions of the very hottest stars (the sun itself is only a middling warm one) may reach as high as 2,000,000,000° K.

Now two billion degrees is a tidy amount of heat even when compared to a muggy day in New York, but the questions arise: How long can this go on? Is there any limit to how hot a thing can be?

Or to put it another way, How hot is hot?

That sounds like asking, How high is up? and I wouldn't do such a thing except that our twentieth century has seen the height of upness scrupulously defied in some respects.

For instance, in the good old days of Newtonian physics there was no recognized limit to velocity. The question, How fast is fast? had no answer. Then along came Einstein, who advanced the notion, now generally accepted, that the maximum possible velocity is that of light, which is equal to 186,274 miles per second, or, in the metric system, 299,776 kilometers per second. *That* is the fastness of fast.

So why not consider the hotness of hot?

One of the reasons I would like to do just that is to take up the question of the various temperature scales and their interconversions for the general edification of the readers. The subject now under discussion affords an excellent opportunity for just that.

For instance, why did I specify the Kelvin scale of temperature in giving the figures above? Would there have been a difference if I had used Fahrenheit? How much and why? Well, let's see.

The measurement of temperature is a modern notion, not more than 350 years old. In order to measure temperature, one must first grasp the idea that there are easily observed physical characteristics which vary more or less uniformly with change in the subjective feeling of "hotness" and "coldness." Once such a characteristic is observed and reduced to quantitative measurement, we can exchange a subjective, "Boy, it's getting hotter," to an objective, "The thermometer has gone up another three degrees."

One applicable physical characteristic, which must have been casually observed by countless people, is the fact that substances expand when warmed and contract when cooled. The first of all those countless people, however, who tried to make use of this fact to measure temperature was the Italian physicist Galileo Galilei. In 1603 he inverted a tube of heated air into a bowl of water. As the air cooled to room temperature, it contracted and drew the water up into the tube. Now Galileo was ready. The water level kept on changing as room temperature changed, being pushed down when it warmed and expanded the trapped air, and being pulled up when it cooled and contracted the trapped air. Galileo had a thermometer (which, in Greek, means "heat measure"). The only trouble was that the basin of water was open to the air and air pressure kept changing. That also shoved the water level up and down, independently of temperature, and queered the results.

By 1654, the Grand Duke of Tuscany, Ferdinand II,

evolved a thermometer that was independent of air pressure. It contained a liquid sealed into a tube, and the contraction and expansion of the liquid itself was used as an indication of temperature change. The volume change in liquids is much smaller than in gases, but by using a sizable reservoir of liquid which was filled so that further expansion could only take place up a very narrow tube, the rise and fall within that tube, for even tiny volume changes, was considerable.

This was the first reasonably accurate thermometer, and was also one of the few occasions when the nobility contributed to scientific advance.

With the development of accuracy, there slowly arose the notion that, instead of just watching the liquid rise and fall in the tube, one ought to mark off the tube at periodic intervals so that an actual quantitative measure could be made. In 1701 Isaac Newton suggested that the thermometer be thrust into melting ice and that the liquid level so obtained be marked as 0, while the level attained at body temperature be marked off as 12, and the interval divided into twelve equal parts.

The use of a twelve-degree scale for this temperature range was logical. The English had a special fondness for the duo-decimal system (and need I say that Newton was English?). There are twelve inches to the foot, twelve ounces to the Troy pound, twelve shillings to the pound, twelve units to a dozen and twelve dozen to a gross. Why not twelve degrees to a temperature range? To try to divide the range into a multiple of twelve degrees—say into twenty-four or thirty-six degrees—would carry the accuracy beyond that which the instrument was then capable of.

But then, in 1714, a German physicist named Gabriel Daniel Fahrenheit made a major step forward. The liquid that had been used in the early thermometers was either water or alcohol. Water, however, froze and became useless at temperatures that were not very cold, while alcohol boiled and became useless at temperatures that were not very hot. What Fahrenheit did was to substitute mercury. Mercury stayed liquid well below the freezing point of water and well above the boiling point of alcohol. Further-more, mercury expanded and contracted more uniformly with temperature than did either water or alcohol. Using mercury, Fahrenheit constructed the best thermometers the world had yet seen.

With his mercury thermometer, Fahrenheit was now ready to use Newton's suggestion; but in doing so, he made a number of modifications. He didn't use the freez-ing point of water for his zero (perhaps because winter temperatures below that point were common enough in Germany and Fahrenheit wanted to avoid the complication of negative temperatures). Instead, he set zero at the very lowest temperature he could get in his laboratory, and that he attained by mixing salt and melting ice.

Then he set human body temperature at 12, following Newton, but that didn't last either. Fahrenheit's thermom-eter was so good that a division into twelve degrees was unnecessarily coarse. Fahrenheit could do eight times as well, so he set body temperature at 96.

On this scale, the freezing point of water stood at a little under 32, and the boiling point at a little under 212. It must have struck him as fortunate that the difference be-tween the two should be about 180 degrees, since 180 was a number that could be divided evenly by a large variety of integers including 2, 3, 4, 5, 6, 9, 10, 12, 15, 18, 20, 30, 36, 45, 60 and 90. Therefore, keeping the zero point as was, Fahrenheit set the freezing point of water at exactly 32 and the boiling point at exactly 212. That made body temperature come out (on the average) at 98.6°, which was an uneven value, but this was a minor point.

Thus was born the Fahrenheit scale, which we, in the United States, use for ordinary purposes to this day. We speak of "degrees Fahrenheit" and symbolize it as "° F." so that the normal body temperature is written 98.6° F.

In 1742, however, the Swedish astronomer Anders Celsius, working with a mercury thermometer, made use of a different scale. He worked downward, setting the boiling point of water equal to zero and the freezing point at 100. The next year this was reversed because of what seems a natural tendency to let numbers increase with increasing heat and not with increasing cold.

Because of the hundredfold division of the temperature range in which water was liquid, this is called the Centigrade scale from Latin words meaning "hundred steps." It is still common to speak of measurements on this scale as "degrees Centigrade," symbolized as "° C." However, a couple of years back, it was decided, at an international conference, to call this scale after the inventor, following the Fahrenheit precedent. Officially, then, one should speak of the "Celsius scale" and of "degrees Celsius." The symbol remains "° C."

The Celsius scale won out in most of the civilized world. Scientists, particularly, found it convenient to bound the liquid range of water by 0° at the freezing end and 100° at the boiling end. Most chemical experiments are conducted in water, and a great many physical experiments, involving heat, make use of water. The liquid range of water is therefore the working range, and as scientists were getting used to forcing measurements into line with the decimal system (soon they were to adopt the metric system which is decimal throughout), 0 and 100 were just right. To divide the range between 0 and 10 would have made the divisions too coarse, and division between 0 and 1000 would have been too fine. But the boundaries of 0 and 100 were just right.

However, the English had adopted the Fahrenheit scale. They stuck with it and passed it on to the colonies which, after becoming the United States of America, stuck with it also.

Of course, part of the English loyalty was the result of their traditional traditionalism, but there was a sensible reason, too. The Fahrenheit scale is peculiarly adapted to meteorology. The extremes of 0 and 100 on the Fahrenheit scale are reasonable extremes of the air temperature in western Europe. To experience temperatures in the shade of less than 0° F. or more than 100° F. would be unusual indeed. The same temperature range is covered on the Celsius scale by the limits −18° C. and 38° C. These are not only uneven figures but include the inconvenience of negative values as well.

So now the Fahrenheit scale is used in English-speaking countries and the Celsius scale everywhere else (including those English-speaking countries that are usually not considered "Anglo-Saxon"). What's more, scientists everywhere, *even* in England and the United States, use the Celsius scale.

If an American is going to get his weather data thrown at him in degrees Fahrenheit and his scientific information in degrees Celsius, it would be nice if he could convert one into the other at will. There are tables and graphs that will do it for him, but one doesn't always carry a little table or graph on one's person. Fortunately, a little arithmetic is all that is really required.

In the first place, the temperature range of liquid water is covered by 180 equal Fahrenheit degrees and also by 100 equal Celsius degrees. From this, we can say at once that 9 Fahrenheit degrees equal 5 Celsius degrees. As a first approximation, we can then say that a number of Celsius degrees multiplied by $\frac{9}{5}$ will give the equivalent number of Fahrenheit degrees. (After all, 5 Celsius degrees multiplied by $\frac{9}{5}$ does indeed give 9 Fahrenheit degrees.)

Now how does this work out in practice? Suppose we are speaking of a temperature of 20° C., meaning by that a temperature that is 20 Celsius degrees above the freezing point of water. If we multiply 20 by $\frac{9}{5}$ we get 36, which is the number of Fahrenheit degrees covering the same range; the range, that is, above the freezing point of water. But the freezing point of water on the Fahrenheit scale is 32°. To say that a temperature is 36 Fahrenheit degrees above the freezing point of water is the same as saying it is 36 plus 32 or 68 Fahrenheit degrees above the Fahrenheit zero; and it is degrees above zero that is signified by the Fahrenheit reading. What we have proved by all this is that 20° C. is the same as 68° F. and vice versa.

This may sound appalling, but you don't have to go through the reasoning each time. All that we have done can be represented in the following equation, where F represents the Fahrenheit reading and C the Celsius reading:

$$F = \frac{9}{5} C + 32 \qquad \text{(Equation 1)}$$

To get an equation that will help you convert a Fahrenheit reading into Celsius with a minimum of thought, it is only necessary to solve Equation 1 for C, and that will give you:

$$C = \frac{9}{5} (F - 32) \qquad \text{(Equation 2)}$$

To give an example of the use of these equations, suppose, for instance, that you know that the boiling point of ethyl alcohol is 78.5° C. at atmospheric pressure and wish to know what the boiling point is on the Fahrenheit scale. You need only substitute 78.5 for C in Equation 1. A little arithmetic and you find your answer to be 173.3° F.

And if you happen to know that normal body temperature is 98.6° F. and want to know the equivalent in Celsius, it is only necessary to substitue 98.6 for F in Equation 2. A little arithmetic again, and the answer is 37.0° C.

But we are not through. In 1787, the French chemist Jacques Alexandre César Charles discovered that when a gas was heated, its volume expanded at a regular rate, and that when it was cooled, its volume contracted at the same rate. This rate was $\frac{1}{273}$ of its volume at 0° C. for each Celsius degree change in temperature.

The expansion of the gas with heat raises no problems, but the contraction gives rise to a curious thought. Suppose a gas has the volume of 273 cubic centimeters at 0° C. and it is cooled. At −1° C. it has lost $\frac{1}{273}$ of its original volume, which comes to 1 cubic centimeter, so that only 272 cubic centimeters are left. At −2° C. it has lost another $\frac{1}{273}$ of its original volume and is down to 271 cubic centimeters. The perceptive reader will see that if this loss of 1 cubic centimeter per degree continues, then at −273° C., the gas will have shrunk to zero volume and will have disappeared from the face of the earth.

Undoubtedly, Charles and those after him realized this, but didn't worry. Gases on cooling do not, in actual fact, follow Charles's law (as this discovery is now called) exactly. The amount of decrease slowly falls off and before the −273° point is reached, all gases (as was guessed then and as is known now) turn to liquids, anyway; and Charles's law does not apply to liquids. Of course, a "perfect gas" may be defined as one for which Charles's law works perfectly. A perfect gas would indeed contract steadily and evenly, would never turn to liquid, and would disappear at −273°. However, since a perfect gas is only a chemist's abstraction and can have no real existence, why worry?

Slowly, through the first half of the nineteenth century, however, gases came to be looked upon as composed of discrete particles called molecules, all of which were in rapid and random motion. The various particles therefore possessed kinetic energy (i.e., "energy of motion"), and temperature came to be looked upon as a measure of the kinetic energy of the molecules of a substance under given conditions. Temperature and kinetic energy rise and fall together. Two substances are at the same temperature when the molecules of each have the same kinetic energy. In fact, it is equality of kinetic energy which our human senses (and our nonhuman thermometers) register as "being of equal temperature."

The individual molecules in a sample of gas do not all possess the same energies, by any means, at any given temperature. There is a large range of energies which are produced by the effect of random collisions that happen to give some molecules large temporary supplies of energy, leaving others with correspondingly little. Over a period of time and distributed among all the molecules present, however, there is an "average kinetic energy" for every temperature, and this is the same for molecules of all substances.

In 1860, the Scottish mathematician Clerk Maxwell worked out equations which expressed the energy distribution of gas molecules at any temperature and gave means of calculating the average kinetic energy.

Shortly after, a British scientist named William Thomson (who had just been raised to the ranks of the nobility with the title of Baron Kelvin) suggested that the kinetic energy of molecules be used to establish a temperature scale. At 0° C. the average kinetic energy per molecule of any substance is some particular value. For each Celsius degree that the temperature is lowered, the molecules lose $\frac{1}{273}$ of their kinetic energy. (This is like Charles's law, but whereas the decrease of gas volume is not perfectly regular, the decrease in molecular energies—of which the decrease in volume is only an unavoidable and imperfect consequence—*is* perfectly regular.) This means that at −273° C., or, more exactly, at −273.16° C., the molecules have zero kinetic energy. The substance—any substance—can be cooled no further, since negative kinetic energy is inconceivable.

The temperature of −273.16° C. can therefore be considered an "absolute zero." If a new scale is now invented in which absolute zero is set equal to 0° and the size of the degree is set equal to that of the ordinary Celsius degree, then any Celsius reading could be converted to a corresponding reading on the new scale by the addition of 273.16. (The new scale is referred to as the absolute scale or, more appropriately in view of the convention that names scales after the inventors, the Kelvin scale, and degrees on this scale can be symbolized as either "° A." or "° K.") Thus, the freezing point of water is 273.16° K. and the boiling point of water is 373.16° K.

In general:

$$K = C + 273.16 \qquad \text{(Equation 3)}$$
$$C = K - 273.16 \qquad \text{(Equation 4)}$$

You might wonder why anyone would need the Kelvin scale. What difference does it make just to add 273.16 to every Celsius reading? What have we gained? Well, a great many physical and chemical properties of matter vary with temperature. To take a simple case, there is the volume of a perfect gas (which is dealt with by Charles's law). The volume of such a gas, at constant pressure, varies with temperature. It would be convenient if we could say that the variation was direct; that is, if doubling the temperature meant doubling the volume.

If, however, we use the Celsius scale, we cannot say this. If we double the temperature from, say, 20° C. to

40° C., the volume of the perfect gas does *not* double. It increases by merely one-eleventh of its original volume. If we use the Kelvin scale, on the other hand, a doubling of temperature does indeed mean a doubling of volume. Raising the temperature from 20° K. to 40° K., then to 160° K. and so on, will double the volume each time.

In short, the Kelvin scale allows us to describe more conveniently the manner in which the universe behaves as temperature is varied—more conveniently than the Celsius scale, or any scale with a zero point anywhere but at absolute zero, can.

Another point I can make here is that in cooling any substance, the physicist is withdrawing kinetic energy from its molecules. Any device ever invented to do this only succeeds in withdrawing a fraction of the kinetic energy present, however little the amount present may be. Less and less energy is left as the withdrawal step is repeated over and over, but the amount left is never zero.

For this reason, scientists have not reached absolute zero and do not expect to, although they have done wonders and reached a temperature of 0.00001° K.

At any rate, here is another limit established, and the question: How cold is cold? is answered.

But the limit of cold is a kind of "depth of down" as far as temperature is concerned, and I'm after the "height of up," the question of whether there is a limit to hotness and, if so, where it might be.

Let's take another look at the kinetic energy of molecules. Elementary physics tells us that the kinetic energy (E) of a moving particle is equal to $\frac{1}{2}mv^2$, where "m" represents the mass of a particle and "v" its velocity. If we solve the equation $E = \frac{1}{2}mv^2$ for "v," we get:

$$v = \sqrt{\frac{2E}{m}} = 1.414 \sqrt{\frac{E}{m}} \quad \text{(Equation 5)}$$

But the kinetic energy content is measured by the temperature (T), as I've already said. Consequently, we can substitute "T" for "E" in Equation 5 (and I will also change the numerical constant to allow the figures to

161

come out correctly in the units I will use). We can then
say that:

$$v = 0.158 \sqrt{\frac{T}{m}}$$ (Equation 6)

Now then, if in Equation 6 the temperature (T) is
given in degrees Kelvin, and the mass (m) of the particle
is given in atomic units, then the average velocity (v) of
the particles will come out in kilometers per second. (If
the numerical constant were changed from 0.158 to
0.098, the answer would come out in miles per second.)

For instance, consider a sample of helium gas. It is
composed of individual helium atoms, each with a mass
of 4, in atomic units. Suppose the temperature of the
sample is the freezing point of water (273° K.) We can
therefore substitute 273 for "T" and 4 for "m" in Equation 6. Working out the arithmetic, we find that the average
velocity of helium atoms at the freezing point of water
is 1.31 kilometers per second (0.81 miles per second).

This will work out for other values of "T" and "m."
The velocity of oxygen molecules (with a mass of 32) at
room temperature (300° K. works out as $0.158 \frac{300}{32}$
or 0.48 kilometers per second). The velocity of carbon
dioxide molecules (with a mass of 44) at the boiling point
of water (373° K.) is 0.46 kilometers per second, and
so on.

Equation 6 tells us that at any given temperature, the
lighter the particle the faster it moves. It also tells us
that at absolute zero (where T = 0) the velocity of any
atom or molecule, whatever its mass, is zero. This is another way of looking at the absoluteness of absolute zero.
It is the point of absolute (well, almost absolute) atomic
or molecular rest.

But if a velocity of zero is a lower limit, is there not an
upper limit to velocity as well? Isn't this upper limit the
velocity of light, as I mentioned at the beginning of the
article? When the temperature goes so high that "v" in
Equation 6 reaches the speed of light and can go no higher,
have we not reached the absolute height of up, the ultimate

hotness of hot? Let's suppose all that is so, and see where it leads us.

Let's begin by solving Equation 6 for "T." It comes out:

$$T = 40 \text{ mv}^2 \qquad \text{(Equation 7)}$$

The factor, 40, only holds when we use units of degrees Kelvin, and kilometers per second.

Let's set the value of "v" (the molecular velocity) equal to the maximum possible, or the 299,776 kilometers per second which is the velocity of light. When we do that, we get what would seem to be the maximum possible temperature (T_{max}):

$$T_{max} = 3,600,000,000,000 \text{ m} \quad \text{(Equation 8)}$$

But now we must know the value of "m" (the mass of the particles involved). The higher the value of "m," the higher the maximum temperature.

Well, at temperatures in the millions all molecules and atoms have broken down to bare nuclei. At temperatures of hundreds of millions and into the low billions, fusion reactions between simple nuclei are possible so that complicated nuclei can be built up. At still higher temperatures, this must be reversed and all nuclei must break apart into simple protons and neutrons.

Let's suppose, then, that in the neighborhood of our maximum possible temperature, which is certainly over a trillion degrees, only protons and neutrons can exist. These have a mass of 1 on the atomic scale. Consequently, from Equation 8, we must conclude that the maximum possible temperature is 3,600,000,000,000° K.

Or must we?

For alas, I must confess that in all my reasoning from Equation 5 on there has been a fallacy. I have assumed that the value of "m" is constant; that if a helium atom has a mass of 4, it has a mass of 4 under all conceivable circumstances. This would be so, as a matter of fact, if the Newtonian view of the universe were correct, but in

163

the Newtonian universe there is no such thing as a maximum velocity and therefore no upper limit to temperature.

On the other hand, the Einsteinian view of the universe, which gives an upper limit of velocity and therefore seems to offer the hope of an upper limit of heat, does not consider mass a constant. The mass of any object (however small under ordinary conditions, as long as it is greater than zero) increases as its velocity increases, becoming indefinitely large as one gets closer and closer to the velocity of light. (A shorthand way of putting this is: "Mass becomes infinite at the velocity of light.") At ordinary velocities, say of no more than a few thousand kilometers per second, the increase in mass is quite small and need not be worried about except in the most refined calculations.

However, when we are working near the velocity of light or even at it as I was trying to do in Equation 8, "m" becomes very large and reaches toward the infinite regardless of the particle being considered, and so consequently does "T_{max}." There is no maximum possible temperature in the Einsteinian universe any more than in the Newtonian. In this particular case, there is no definite height to up.

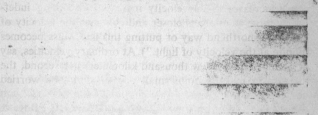

IV ASTRONOMY

13 Hot Stuff

It's the life's ambition of every decent, right-thinking scientist or near-scientist (I use the latter noun as an excuse to include myself) to influence the course of science. For the better, of course.

Most of us, alas, have to give up that ambition; I did so long ago. Never (so my heart told me) would there be an "Asimov's law" to brighten the pages of a physics textbook, or an "Asimov reaction" to do the same for those of a chemistry textbook. Slowly, the possibility of an "Asimov theory" and even an "Asimov conjecture" slipped through my fingers, and I was left with nothing.

With nothing, that is, but my electric typewriter and my big mouth, and the hidden hope that some idle speculation of my own might spark better minds than mine into some worthwhile accomplishment.

Well, it's happened.

And here's how it happened.

Some weeks after the material in the previous chapter was first published, I received a letter from a post-doctoral research worker at the Institute for Advanced Study at Princeton—a gentleman by the name of Hong Yee Chiu.

He gave me his own thoughts on the maximum possible temperature, pointing out that my own results arose out of the assumption that the universe was infinite. If the universe were finite, then it had a finite mass. If that finite mass (but for one particle) were converted completely into energy and that energy were concentrated in the one remaining particle, and if we pretend that temperature has meaning in systems consisting of but one particle, then we would end with the maximum conceivable temperature for the actual universe. He calculated what that temperature would be. It came out to a tremendously high, but, of course, not infinite, temperature.

However, the problem of the maximum possible temperature, under the actual conditions of the universe, continued to occupy his thinking, even after he left Princeton and took a position with the Institute of Space Studies in New York. According to a letter I received from him, dated November 14, 1961 (which he kindly gave me permission to quote):

"I switched from the field of elementary particle physics to astrophysics then, right after I got my degree. Your article initiated my interest in the field of supernova. As one knows, the interior of a star is hotter than anything one can think of. Will there be an upper limit for temperature there?"

The result of his thinking appeared in papers in *Physical Reviews* and *Annals of Physics* outlining a new theory of supernova formation.

I would like, out of sheer proprietary interest, to give you some notion of this new theory but, please note, I hereby absolve Dr. Chiu of any responsibility for what I say. In his papers, you see, he uses double integrals and hyperbolic functions and all sorts of mathematical devices that are slightly beyond the range of elementary algebra and that rather leave me at loose ends. Consequently, I may misunderstand some of the things he says.

However, I have done what I can and, as I always do, I will begin at the beginning.

The beginning is the neutrino, a subatomic particle with a fascinating history that goes back to Einstein. In 1905, in his Special Theory of Relativity, Einstein pointed out that mass was a form of energy, and that its energy value could be calculated by a simple formula. (Yes, I'm referring to $E = MC^2$.)

This formula was applied to alpha particle production, for instance. The uranium atom lost an alpha particle and became a thorium atom. The alpha particle and the thorium atom together had a mass just slightly less than that of the original uranium atom. This mass had not disappeared; it had been converted into the kinetic energy of the speeding alpha particle. Consequently, all the alpha particles produced by a given type of atom had the same energy content. (Or, rather, one of a small number of different energy contents, for a given type of atom can exist at several different energy levels; and, at a higher energy level, it will give off a somewhat more energetic alpha particle.)

All this was very satisfactory, of course. Mass was converted into energy and the ledgers balanced and physicists rubbed their hands gleefully. The next step was to show that the energy ledgers balanced in the case of beta particles production, too. To be sure a beta particle (an electron) was only $\frac{1}{7,350}$ as massive as an alpha particle (a helium nucleus), but that shouldn't affect anything. The principle was the same.

And yet isotopes that emitted beta particles did *not* emit them all at the same energies, or at a small number of specific different energies.

What actually turned out to be the case was that beta particles were given off at *any* energy up to a certain maximum. The maximum energy was that which just accounted for the loss of mass, but only a vanishingly small number of electrons attained this. Virtually all the particles came off at smaller energies and some came off at very small energies indeed.

The net result was that there was some energy missing, and the ledgers did not balance.

This created, you might well imagine, a certain amount of lip-biting and brow-furrowing among physicists; at least as much as it would have among bank-examiners. After all, if energy were really disappearing, then the law of conservation of energy was broken, and no physicist in his right mind wanted to allow that to happen until every other conceivable alternative had been explored.

In 1931, the Austrian-born physicist Wolfgang Pauli came up with a suggestion. If the electron were not carrying off all the energy that was available through mass-loss, then another particle must be. This other particle, however, went undetected, so it must lack detectable properties. Of these undetectable properties, electric charge was foremost; consequently Pauli postulated a neutral particle.

Furthermore, the amount of kinetic energy left over in beta-particle production was not enough to make a very large particle if it were reconverted to mass, especially since much of the energy had to be energy of motion. The particle was certain to have only a fraction of the mass of even an electron and it was quite likely that it had no mass at all.

Pauli's suggested particle was the nearest thing to nothing one could imagine.

No charge, no mass—just a speeding ghost of a particle which carried off the energy that would otherwise be left unaccounted for.

In 1932, a heavy neutral particle (as heavy as the proton) was detected and named "neutron." Consequently, the Italian physicist Enrico Fermi suggested that Pauli's particle, neutral but much smaller than the neutron, be called "neutrino" (Italian for "little neutral one").

The neutrino turned out to be remarkably useful. Not only did it save the law of conservation of energy, but also the law of conservation of particle spin and of particle-antiparticle production.

But was it any more than a Finagle's Constant designed to make a wrong answer right? Did the neutrino really exist or was it just an *ad hoc* device, invented by the agile minds of physicists to keep their rickety structure of supposed reality standing?

The situation could be resolved if the neutrino were only detected, actually detected. Yet, to be detected, it had to interact with other particles. But, unfortunately, neutrinos did not interact with other particles; or at least they interacted so rarely that it seemed scarcely worth talking about.

It was calculated that a neutrino could travel through one hundred light-years of water before there was as much as a fifty-fifty chance of interaction; and, as you can well imagine, it is hard to set up a tube of water a hundred light-years long.

However, there is a 25 per cent chance that it will interact after passing through only light-years of water, and a 12.5 per cent chance that it will interact after passing through a mere twenty-five light-years of water, and so on. In fact, there is a terribly small, *but finite,* chance that it will interact after passing through, say, six feet of water.

The chance is so small, however, that to wait for one neutrino to do so is foolish. But why work with just one neutrino? A nuclear reactor is constantly liberating vast quantities of neutrinos (if they exist). If one places tanks of water near a nuclear reactor and sets up devices that will detect gamma radiation of just the wave length to be expected when a neutrino interacts with a proton, the chance that one out of a vast number of neutrinos will interact within a few feet, becomes rather decent.

And, as a matter of fact, in 1953 this was accomplished at Los Alamos, and the existence of the neutrino was proven. It wasn't a Finagle's Constant at all, but a real live particle. It had no mass and no charge and was still the nearest thing to nothing you could imagine, but *it was there,* and that's what counts.

When is a neutrino produced? The best known neutrino-producing reactions are those involving neutron-proton interchanges. When a neutron is converted into a proton and an electron, a neutrino is produced. When a proton is converted into a neutron and a positron, an antineutrino is produced. (The neutrino and antineutrino are distinct particles, differing in spin, but both are massless and chargeless, and for purposes of this article I shall lump them both together as neutrinos.)

By far the largest producers of neutrinos are the stars. Consider the sun, for instance. Its power is derived from the conversion of hydrogen to helium. The hydrogen nucleus is a single proton while the helium nucleus consists of two protons and two neutrons. In converting four hydrogen nuclei to one helium nucleus, therefore, two of the four protons of the hydrogen nuclei must be converted to neutrons, with the production of two neutrinos (and other things, too, like positrons and photons). For every two hydrogen atoms consumed, then, one neutrino is formed.

In order to maintain its energy output, the sun must convert 4,200,000 tons of mass into energy every second. In converting hydrogen to helium, 0.75 per cent of the mass is lost, so that in order to lose 4,200,000 tons, 560,000,000 tons of hydrogen must be processed.

Losing over half a billion tons of hydrogen every second sounds like a fearsome loss, but don't worry. Some three-fifths of the mass of the sun is hydrogen, so that there is well over an octillion tons of hydrogen available in the sun. If hydrogen continued to be consumed at the present rate, and no other nuclear processes were involved, the hydrogen content of the sun would last for some 60,000,000,000 years. And you and I would very likely be dead by then.

Anyway, the conversion of 560,000,000 tons of hydrogen per second means that 2.8×10^{38} hydrogen atoms must be fed into the rentless maw of the sun's nuclear engine every second. Therefore, 1.4×10^{38} neutrinos are produced each second.

The neutrinos, produced in the sun's interior, radiate outward in all directions. Naturally, almost all of them miss the tiny target presented by the earth, nearly 93,000,000 miles from the sun. However, it has been estimated (according to some work quoted in one of Dr. Chiu's papers) that even on earth, nearly 10,000,000,000 neutrinos from the sun are passing through every square centimeter of cross-section.

This means that they pass through the atmosphere, through the oceans, through the crust, through the central core, *through you*. They pass through you constantly, whether the sky is cloudy or clear and whether it is day or night. If it is night, the neutrinos pass through the body of the earth and get you anyway. And since they travel at the speed of light, the delay in their getting you, over the added distance of the earth's diameter, is only about $\frac{1}{28}$ of a second.

Again, fear not. Let us say that you are constantly presenting the maximum surface to neutrino's bombardment and that this represents 10,000 square centimeters (which is generous). One hundred trillion (100,000,000,-000,000) neutrinos would then be passing through you every second.

You are mostly water and one neutrino must pass through a hundred light-years of water for a fifty-fifty chance of reaction. However, when you expose maximum surface to the neutrinos, you are only about a foot thick. All hundred trillion neutrinos pass through a total of a hundred trillion feet of water, or roughly $\frac{1}{300}$ a light-year. This means that on the average one neutrino will react with a particle within your body every 30,000 seconds (there is a fifty-fifty chance of it, anyway), or once just about every eight hours, while several quintillion neutrinos pass through you in lofty indifference.

And one neutrino interaction every eight hours?

Why, in a single minute, 1,200,000 atoms of potassium-40 and 180,000 atoms of carbon-14 (both naturally present within the body and both naturally radioactive) break down, spraying the body with beta particles and radiation.

So forget the neutrinos.

The interior of the sun is at a temperature of perhaps 20,000,000 degrees. (Throughout this article I shall mean Centigrade when I talk of degrees.) It has to be that high to produce enough expansive force, through radiation pressure and through the kinetic energy of particles, to counteract the enormous pressures produced by gravitation which tend to contract the sun.

A star exhibits this same tug of war. Mass (and consequent gravitational force) tends to contract it; temperature (and consequent radiational force) tends to expand it. As long as the two are in reasonable balance, all is well.

As hydrogen is converted to helium, however, the four protons of hydrogen, comparatively loosely packed to begin with, are converted into the compact two-proton and two-neutron arrangement of the helium nucleus. The density of the star's center increases and, as more and more helium is formed, the concentration of mass and, consequently, the intensity of the gravitational field, increases. To counteract this and to retain equilibrium, the temperature of the star's center rises.

At some point, eventually, the temperature rises high enough to ignite the helium nuclei, forcing them into fusion reactions that form still more complex nuclei. This process continues, with temperature steadily increasing, so that successively more complex atoms are produced. Finally, iron atoms are produced.

The iron atoms are about the most complicated that can be formed by ordinary stellar reactions. No further increases in nuclear complexity will produce energy. Atoms more complicated than iron require an *input* of energy. Iron, therefore, represents the dead end of ordinary stellar life.

At this dead end, the star has the aspect of an onion, consisting of layers of different chemical composition. At the very center is the iron core, surrounded by a silicon layer, surrounded by a magnesium layer, surrounded by a neon layer, surrounded by a carbon layer, surrounded by a helium layer, surrounded by a hydrogen layer that forms the surface of the star.

Each layer constantly undergoes fusion reactions producing heavier nuclei that are added to the layer beneath, with the iron core the net gainer and the hydrogen surface the net loser. The gravitation field continues to increase, but now there is no additional energy formation possible at the center to balance it.

As the center continues to heat up, some crucial line is passed and the star suddenly collapses. In so collapsing, the sudden increase in pressure on the outer layers, where fusable fuel still exists, hastens the fusion reactions, producing a vast outflux of energy that succeeds in blowing the star to kingdom come.

The result is a huge supernova (see "The Sight of Home," *Fact and Fancy,* Doubleday, 1962) out of the energy of which even iron atoms are fused to produce still more complicated atoms—all the way up to uranium at least, and very likely up to californium. The explosion spreads these heavy elements out into space, and new stars and stellar systems are formed (like our own) that will possess small quantities to begin with.

Does this mean that every star is fated to become a supernova at some later stage of its life? Apparently not.

The more massive a star, the more intense its gravitational field and therefore the higher its internal temperature and the greater its luminosity at a given stage of its nuclear-reaction cycle. (This is the "mass-luminosity law," announced by the English astronomer Arthur S. Eddington in 1924. He was the first to calculate the enormous central temperatures of stars.) Apparently, in order for a star to reach the point where a supernova explosion is set off, it must start with at least 1.5 times the mass of our sun. This is called "Chandrasekhar's limit" after the astronomer who first worked it out. So whatever will happen to our sun, it will never go supernova. It can't ever get hot enough.

But what is the exact nuclear process that leads to this spectacular collapse and explosion? And, in particular, what about the actual temperatures at the center of a star about to go supernova? That certainly is the highest temperature actually reached in the universe, and *that* is what Dr. Chiu is after.

Well, stars get rid of energy in two ways. They produce electromagnetic radiation and they produce neutrinos. The behavior of the two is different. Electromagnetic radiation interacts strongly with matter so that the gamma rays produced at the center of the sun are forever striking protons and neutrons and alpha particles and being absorbed and reemitted and so on. It is a long and tedious process for radiation to make its way out from the center of the sun to its surface.

The best indication of this is that the temperature of the sun's surface is a mere 6,000 degrees. You may consider this hot, and by earthly standards it *is* hot. Still, remember that the sun's surface is only 400,000 miles away from a large body of matter that is at a temperature of 20,000,000 degrees. If there were nothing between the sun's core and a point 400,000 miles away, any matter at that point would itself be at a temperature of millions of degrees. For matter at that point to be at a mere 6,000 degrees indicates what a superlatively excellent insulator the substance of the sun is and how difficult it is for radiation to get through that substance and to escape into space.

The energy carried off by the neutrinos, however, behaves differently. Neutrinos simply streak out of the sun's center, where they are produced, at the speed of light. They completely disregard the ordinary matter of the sun and are beyond its substance in less than three seconds.

In the case of the sun, however, the fraction of the total energy that escapes as neutrinos is rather small. The energy loss by neutrino escape results in some slight cooling of the solar interior, of course, but that is made up for by a slight contraction (slight enough to be indetectable) of the sun.

And in the stages beyond helium, neutrino production becomes even more unimportant if proton-neutron interchanges alone are taken into account.

Thus, suppose we begin with 56 hydrogen nuclei. These are converted into 19 helium nuclei, which, in the later stages of stellar life, are in turn finally converted into a single iron nucleus.

The 56 hydrogen nuclei consist of 56 protons.

The 19 helium nuclei consist of 28 protons and 28 neutrons, segregated into groups of two-proton plus two-neutron units.

The one iron atom consists of 26 protons and 30 neutrons, all crowded into a single nucleus.

In going from hydrogen to helium, then, 28 protons must be converted to 28 neutrons with the production of 28 neutrinos.

In going from helium to iron, only 2 more protons need be converted to neutrons, with the production of only 2 neutrinos.

It would seem, then, that except in the initial hydrogen-to-helium stage, neutrino production can be ignored; and since it plays little role in the functioning of the sun where hydrogen-to-helium is the big thing, it should most certainly play little role in the function of stars advanced into helium-burning and beyond.

This is where Dr. Chiu's new theories come in. Dr. Chiu suggests two new manners in which neutrinos can be formed. He suggests that electromagnetic radiation itself may interact to form neutrinos. In addition an electron and a positron may interact to form them.

These reactions happen so rarely at low temperatures, such as the miserable 20,000,000 degrees of the sun's interior, that neutrino formation by the Chiu reactions can be ignored. As the temperature rises, however, this formation becomes increasingly important.

By the time a temperature of one or two billion degrees (the temperature required for the formation of iron nuclei) is reached, neutrino formation by the Chiu reactions is considerably more rapid than neutrino formation from proton-neutron interchanges.

This means that a sizable portion of the star's radiation, which, as radiation, can only escape from the star with excessive slowness, is, instead, converted to neutrinos which go zip! and are gone. Nevertheless, the star can still, albeit with difficulty, replace the lost energy by non-catastrophic contraction.

By the time a temperature of 6,000,000,000 degrees is reached, however, neutrinos are formed at such a rate that the heat of the vast stellar interior is carried off in a matter of fifteen or twenty minutes and the star collapses!

Whoosh! There's your supernova.

This means that 6,000,000,000 degrees is the practical upper limit of temperature that can be built up in this universe. The true hot stuff of the universe is the material at the center of the stars, and this can never reach 6,000,-000,000 degrees without initiating an explosion that cools it off. The question I posed in the previous chapter is thus answered.

Dr. Chiu goes on to suggest that if this theory is correct, it ought to be possible to detect stars that are about to go supernova by the quantity of neutrinos they put out. As supernovahood is approached, the rate, according to Dr. Chiu, reaches 10^{53} per second. This is a quadrillion times as many as the sun is producing. Even at a distance of one hundred light years, the number of neutrinos reaching an observer from the direction of the potential supernova is at least a thousand times greater than that which reaches us from the sun.

"Therefore," says Dr. Chiu, in one of his papers, "the establishment of a neutrino monitor station in terrestrial or spatial laboratories may help us predict forthcoming supernovae."

So there you are!

Now, I may be prejudiced, but I think this theory makes so much sense that it will be adopted and praised by all astronomers. And when Dr. Chiu achieves the world-wide fame that I can see now should be his by right, I can hug myself with the pleasant knowledge that an article of mine started it all.

Of course, no one will know this except Dr. Chiu and myself—and the readers of this chapter—and strangers I intend to buttonhole in the street—and people who listen to the television spot-announcements I intend to purchase —and . . .

14 Recipe for a Planet

Slowly, American scientists (and, I believe, Soviet scientists, independently) are making ready to drill a hole through the earth's crust to reach the layer beneath.

This projected "Mohole" (and I'll explain the name, for those of you who happen not to know, later on) will, if it succeeds, bring us the first direct information concerning any portion of our planet other than the very rind. This is exciting for several reasons, one of which is that it will lower the high-blood pressure of many a geologist who for years has had to watch man make ready to go millions of miles out in space while totally unable to penetrate more than a few miles below earth's outer surface. And there is something annoying (if you're a geologist) in the thought that mankind will certainly feel, in its own corporeal hands, a sample of the surface of Mars long before it can possibly feel a sample from the central regions of our own planet.

And yet we ought to look at the bright side. The wonder is not that we are so helpless in the face of some thousands of miles of rigid impenetrability. Naturally, we're helpless. The wonder is that, being so helpless, we have nevertheless deduced as much information about the interior of the earth as we have.

Of course, there are parts of the earth that we can see and feel and which we can subject to our various instruments. Once modern chemistry was established by Lavoisier, there was no serious trouble in analyzing the composition of the atmosphere and of the oceans ("hydrosphere"). The former is, essentially, a mixture of oxygen, nitrogen and argon gases in the ratio, roughly, of 78 : 21 : 1. The latter is, essentially, a 3 per cent water solution of sodium chloride, with some added impurities.

In addition, the outermost portions of the solid matter forming the body of the planet ("lithosphere") are within reach. This, however, presents a new problem. Atmosphere and hydrosphere are homogeneous; that is, if you analyze any small portion of it, you have the composition of the whole. The solid earth itself is heterogeneous; one portion is not necessarily at all like another, which is why we have diamonds in Kimberley and gold in the Klondike but nothing but cheap dirt and some dowdy crabgrass in my backyard.

This means that in order to find out the overall composition of the soil and rocks, analyses have to be run on different samples from different areas of the world, and some sort of average must be taken after estimating that there is so much of this kind of rock on earth and so much of that. Various geologists have done this and come up with estimates that agree fairly well.

A typical such estimate is presented here, with the major elements of the earth's crust presented in order of percentage by weight:

oxygen	46.20
silicon	27.72
aluminum	8.13
iron	5.00
calcium	3.63
sodium	2.83
potassium	2.59
magnesium	2.09

The eight elements make up just over 98.5 per cent of the weight of the earth's outermost layer. The remaining ninety-odd elements can be considered as trace impurities (very important ones in some cases, to be sure, since included among them are elements such as carbon, hydrogen, nitrogen and phosphorus, which are essential to life).

Now none of the elements in the list occurs free; all are found in combination—with each other, naturally, since there is little else to combine with. The most obvious combination is that between silicon and oxygen (which together

178

make up three-fourths of the weight of earth's outermost layer) to form silicon dioxide or silica. Quartz is an example of relatively pure silica, while flint is less pure. Sand is weathered silica. In combination with the other six elements listed (all metals), silicon and oxygen form silicates.

In brief, then, the solid earth's reachable portion can be looked on as a mixture of silica and silicates, with all else chicken feed, at least in terms of quantity.

The distribution of elements in the earth's crust seems lopsided, but, as it happens, when we calculate that distribution by weight, as in the above list, we are making it as unlopsided as possible. Let us suppose that we estimated the composition by numbers of atoms instead of by weight.

Of the eight major elements of the earth's crust, oxygen happens to have the lightest atom. That means that a fixed weight of oxygen will contain 1.75 times as many atoms as that same weight of silicon, 2.5 times as many as the same weight of potassium, 3.5 times as many as the same weight of iron.

If you count by atom, then, it turns out that of every 100 atoms in the earth's crust, 62.5 are oxygen. To put it another way, pick up a handful of soil and chances are that five out of eight of the atoms you are holding are oxygen.

But matters are even more lopsided than that. In forming compounds with silicon and with the six major metals, the oxygen atom accepts electrons; the others all donate them. When an atom accepts electrons, those additional electrons take up orbits (to use the term loosely) on the very outskirts of the atom, swooping far out from the nucleus, which holds them rather weakly. Since the radius of an anion (i.e. an atom plus one or more electrons in excess) extends to the farthest electronic orbit, the oxygen anion is larger than the oxygen atom proper.

On the other hand, an element that gives up an electron or two, gives up just those outermost ones that are most weakly held. The remaining electrons cluster relatively closely about the nucleus, and the radius of such a cation (i.e. an atom with a deficiency of one or more electrons) is smaller than that of the original atom.

The result is that the oxygen anion has a radius of 1.40 angstroms (an angstrom unit is a hundred-millionth of a centimeter), while the silicon cation has a radius of 0.42 angstroms and the iron cation has one of 0.74 angstroms. This despite the fact that the silicon and iron cations are each considerably more massive than the relatively light oxygen anion.

The volume of any sphere varies as the cube of the radius, so that the discrepancy in radii among the ions becomes much magnified in the volume itself. For instance, the volume of the oxygen anion is about 11.5 cubic angstroms, while the volume of the iron cation is only 2.1 cubic angstroms and the volume of the silicon cation is only 0.4 cubic angstroms.

Allowing for the greater number of oxygen atoms and the greater volume of the individual oxygen anion, it turns out that no less than 93.77 per cent of the *volume* of the earth's crust is taken up by oxygen. The solid earth on which we walk is a well-packed set of oxygen anions, crowded closely together, with the small cations of the other seven elements tucked in here and there in the interstices.

That goes for the Rock of Gibraltar, too—just a heap of oxygen and little more.

All this data deal, of course, with those portions of the lithosphere which we can gouge out and pulverize and put through the analytic mill. What about those portions we can't test? Mankind has dug some three miles deep into the crust in pursuit of gold, and a couple of miles deeper in chase of oil, but these are highly localized pinpricks. All but the surface is beyond our ken and may even be forever beyond said ken.

The lazy man's solution to the problem is to suppose that, in general, the surface of the earth's crust is a fair representation of its interior and that the planet is the same through and through as it is on the surface.

Unfortunately for those seeking a simple answer, this isn't so on the face of it. If the earth as a whole were as rich in uranium and thorium as the crust is, our planet would melt with the quantity of radioactive heat radiated. Just the fact that the earth is solid, then, shows that those

180

two elements peter out a short distance below earth's skin, and proves that in one small way, at least, heterogeneity with depth exists.

Furthermore, the predominant rock of the continental masses is granite, while the predominant rock of the ocean bottoms would seem to be basalt. Granite is richer in aluminum that basalt is, and poorer in magnesium, so that some geologists have visualized the earth's crust as consisting of comparatively light continental blocks rich in aluminum silicate ("sial") floating on a comparatively heavy underpinning rich in magnesium silicate ("sima"), with the earth's water supply filling the gaps between the sial blocks.

This is probably an oversimplification, but it still brings up the notion that the composition of the earth changes with depth. And yet, so far, it is only the metals that are involved. There is nothing in what I have said that seems to affect the point of silicon and oxygen preponderance. Whatever the change in detail, the earth might still be a silicate ball in essence—one big globe of rock, in other words.

The first actual information, as opposed to pure guesswork, that was obtained about the earth's deep interior, came when Henry Cavendish first determined the mass of the planet in 1798. The volume was known since ancient Greek times; and dividing Cavendish's mass by the volume gave the overall density of the earth as 5.52 grams per cubic centimeter.

Now, the density of the earth's crust is about 2.8 grams per cubic centimeter, and this means that the density must rise with increasing depth. In fact, it must rise well beyond the 5.52 mark to make up for the lower-than-average density of the surface layers.

This, in itself, is no blow to the rock-ball theory of earth's structure, because, obviously, pressure must increase with depth. The weight of overlying layers of rocks must compress the lower layers more and more down to the center where, it is estimated, the pressure is something like 50,000,000 pounds to a square inch. The same rock which had a density of 2.8 grams per cubic centimeter on the surface might conceivably be squeezed to perdition and a density of, say, 12 grams per cubic centimeter at the earth's center.

A more direct line of attack on the deep interior involves the study of earthquakes. By 1900, the earth was beginning to be girdled by a network of seismographs equipped to study the vibrations set up in the body of the planet by the quakes.

Two main types of earthquake waves are produced, the P (or primary) waves and the S (or secondary) waves. The P waves are longitudinal alternating bands of compression and expansion, like sound waves. The S waves are transverse and have the ordinary snakelike wiggles we associate with waves. The P waves travel more rapidly than do the S waves, and are the first to arrive at a station. The further a station from the actual earthquake, the greater the lag in time before the S waves arrive. Three stations working together can use such time-lag data to spot, with great precision, the point of origin ("epicenter") of the earthquake.

Knowing the location of both the earthquake and the station, it is possible to plot the general path taken by the waves through the body of the earth. The greater the distance between the earthquake and the receiving station, the more deeply would the arriving waves have penetrated the body of the earth. If the earth were uniformly dense and rigid, the time taken by the waves to arrive would be in proportion to the distance traveled.

Actually, however, the density of the earth's substance is not uniform with depth, and neither is the rigidity of the material. Laboratory experiments on various rocks have shown how the velocity of the two types of waves vary with differing degrees of density and rigidity under various temperatures and pressures. Such data can be extrapolated to levels of temperature and pressure that are encountered in the depths of the earth but cannot be duplicated in the laboratory. This is admittedly a risky business—extrapolation always is—but geologists feel confident they can translate the actual velocity of earthquake waves at a given depth into the density of the rock at that depth.

It turns out, then, that the density of the earth does increase fairly slowly and smoothly from the 2.8 grams per cubic centimeter at the surface to about 5.9 grams per cubic centimeter at a depth of some 2,150 miles.

And then, suddenly, there is a sharp break. The fact that this is so can again be told from the behavior of the earthquake waves. As the waves progress through regions of increasing density with depth, then decreasing density as they approach the surface again, they change direction and are refracted, just as light waves are refracted on passing through changing densities of air. As long as the density change is gradual, the direction change is gradual too, and forms a smooth curve. This is exactly what happens as long as the wave in its progress does not penetrate more than 2,150 miles beneath the surface.

Imagine a station, then, at such a distance from the quake itself that the resulting waves have penetrated to that depth. All stations between itself and the quake receive waves, too, at times varying with the distance, with penetrations of less than 2,150 miles.

A station somewhat further from the earthquake epicenter might expect, reasonably enough, to receive waves that have penetrated deeper than 2,150 miles, but it receives no waves at all. Yet stations still further away by a thousand miles or so may receive waves clearly, albeit after a longer interval.

In short, there is an area (the "shadow zone") over the earth's surface, forming a kind of doughnut-shape at a fixed distance from each particular earthquake epicenter, in which no waves are felt. The interpretation is that there is a sudden sharp change in direction in any wave penetrating past the 2,150 mile mark, so that it is sent beyond the shadow zone. The only reason for such a sudden sharp change in direction would be a sudden sharp change in density.

Analysis of arrival times outside the shadow zone shows that this sharp rise in density is from 5.9 to 9.5 grams per cubic centimeter. Below 2,150 miles, the density continues to rise smoothly with increasing depth, reaching a value of about 12 grams per cubic centimeter at the center of the earth.

All this refers to the P waves only. The S waves are even more dramatic in their behavior. When an S wave penetrates below 2,150 miles, it is not merely altered in

183

direction—it is stopped altogether. The most logical explanation arises from the known fact that longitudinal waves such as the P waves can travel through the body of a liquid, while transverse waves, such as the S waves, cannot. Therefore, the regions of the earth lower than 2,150 miles must be liquid.

On the basis of earthquake wave data, then, we must assume the earth to consist of a liquid "core" about 1,800 miles in radius, which is surrounded by a solid "mantle" about 2,150 miles thick. The sharp division between these two major portions was the first clearly demonstrated by the work of the American geologist Beno Gutenberg in 1914, and is therefore called the "Gutenberg discontinuity."

In 1909, a Serbian geologist named Andrija Mohorovicic (there are two little accent marks, different ones, over the two c's in his name which for simplicity's sake, I will omit) discovered a sudden change in the velocity of earthquake waves on a line about 20 miles below the surface. This is called the "Mohorovicic discontinuity" and, because imagination boggles at the numbers of tonsils that would be thrown into disorder every time such a phrase was mouthed, it is becoming customary to say "Moho" instead. Moho is taken as the line separating the mantle below from the "crust" above.

Since those days, a detailed study of Moho shows that it is not at uniform depth. Under coastal land areas the depth *is* about 20 miles (it is 22 miles under New York, for instance), but under mountainous areas it dives lower—as low as 40 miles. (Since the crust is lighter than the mantle, you could say that mountains are mountains because the unusual concentration of light crust present causes the region to "float high.")

Conversely, Moho comes fairly close to the surface under some parts of the dense ocean bottom, which being comparatively heavy for crust, "float low." In some places, Moho is only 8 to 10 miles below the sea level. This is particularly interesting because the ocean itself is 5 to 7 miles deep in spots and there is no problem in drilling through water. The actual thickness of solid material between ourselves and Moho can be as little as 3 miles if the right spot in the ocean is selected.

184

In one of those spots we propose to dig what you can now see must be called the "Mohole" (what else?) to reach the mantle. What's more, my own suggestion is that the ship carrying the equipment be named the "Moholder," but I guess that no one will pay any mind to me at all, with regard to this.

If we are going to consider the overall composition of the earth, it is only necessary to deal with the core and the mantle. The core makes up only one-sixth of the earth's volume, but because of its relatively high density it represents almost one-third of its mass. The remaining two-thirds is the mantle. The crust makes up only $\frac{1}{250}$ of the earth's mass, and the hydrosphere and atmosphere are even more significant than that. We can therefore completely ignore the only parts of the earth on which we have direct analytical data.

What about the mantle and core, then? The mantle differs from the crust only slightly in density and other properties, and everyone agrees that it must be essentially silicate in nature. Experiments in the laboratory show that a rock called olivine (a magnesium iron silicate) will, under high pressure, carry vibrations in the same range of velocities as that at which the mantle transmits earthquake waves. The general impression then is that the mantle differs from the crust in being much more homogeneous, richer in magnesium and poorer in aluminum.

And the core? Still silicate, perhaps, but silicate which, at 2,150 miles, undergoes a sudden change in structure? In other words, may it not be that silicate is squeezed more and more tightly with mounting pressure until, at a certain point, something suddenly gives and all the atoms move into a far more compact arrangement? (This is analogous to the way in which the carbon atoms of graphite will move into the more compact arrangement of diamond if pressure and temperature are high enough.)

Some have proposed this, but there is no actual evidence that at the pressures and temperatures involved (which cannot be duplicated as yet in the laboratory) silicate will behave so.

The alternative is that there is a sudden change in chemical nature of the earth's body, with the comparatively light silicate of the mantle giving way to some substance, heavier and liquid, that will make up the core.

But what material? If we arbitrarily restrict ourselves only to the elements common in the crust, the only substance that would be dense enough at deep-earth pressures (and not too dense) and liquid at deep-earth temperatures would be iron.

But isn't this pulling a rabbit out of the hat, too?

Not quite. There is still another line of evidence that, while terribly indirect, is very dramatic. The iron core was first suggested by a French geologist named Daubrée in 1866, a generation before earthquake data had pinpointed the existence of a core of some sort. His reasoning was based on the fact that so many meteorites consisted almost entirely of iron. This meant that pure iron could be expected to make up a portion of astronomical bodies. Hence, why not an iron core to earth?

As a matter of fact, there are three kinds of meteorites: the "iron meteorites" already referred to; a group of much more common "stony meteorites" and a relatively rare group of "troilite meteorites." It is almost overwhelmingly tempting to suppose that these meteorites are remnants of an earthlike planet (between Mars and Jupiter, where else?) that broke into fragments; that the stony meteorites are fragments of the mantle of that planet; the iron meteorites fragments of its core; and the troilite meteorites fragments of an intermediate zone at the bottom of its mantle.

If this is so, and most geologists seem to assume it is, then by analyzing the three sets of meteorites, we are, in effect, analyzing the earth's mantle and core, at least roughly.

The stony meteorites have, on the whole, the following composition in per cent by weight:

oxygen	43.12
silicon	21.61
magnesium	16.62
iron	13.23
calcium	2.07
aluminum	1.83

As you see, the stony meteorites are mainly a magnesium iron silicate, essentially an olivine. This, with calcium and aluminum as major impurities, comes to 98.5 per cent of the whole. Sodium and potassium, the other metals common in the crust, apparently peter out somewhat in the mantle. It's good to have them where we can reach them, though. They're useful elements and essential to life.

The composition of iron meteorites by per cent looks like this:

iron	90.78
nickel	8.59
cobalt	0.63

Nothing else is present in significant quantity. Because of such analyses, the earth's core is often referred to as the "nickel-iron core."

As for the composition of the troilite meteorites, it is:

iron	61.1
sulfur	34.3
nickel	2.9

This group consists, essentially, of iron sulfide* with a nickel sulfide impurity. Geologists therefore feel that the lowest portion of the earth's mantle may well consist of a zone of iron sulfide making up perhaps one-twelfth of the earth's total mass.

In order to get a notion of the overall composition of the earth, under the assumption that its various major divisions correspond to the different varieties of meteorites, it is only necessary to get a properly weighted average of the meteoric data. Various geologists have made somewhat different assumptions as to the general composition of this or that part of the mantle and have come up with tables that differ in detail but agree in general.

---

*"Troilite" is the name given to an iron sulfide mineral similar to that found in these meteorites. It is named for an eighteenth-century Italian, Domenico Troili.

Here, then, is one summary of the chemical composition of the whole earth in percentages by weight:

iron	35.4
oxygen	27.8
magnesium	17.0
silicon	12.6
sulfur	2.7
nickel	2.7

These six elements make up 98 per cent of the entire globe. If, however, the elements were listed not by weight but by atom number, the relatively light oxygen atoms would gain at the expense of the others and would move into first place. In fact, nearly half (47.2 per cent) of all the atoms in the earth are oxygen.

It remains now to give the recipe for a planet such as ours; and I imagine it ought to run something like this, as it would appear in *Mother Stellar's Planetary Cookbook:*

"Weigh out roughly two septillion kilograms of iron, adding 10 per cent of nickel as stiffening. Mix well with four septillion kilograms of magnesium silicate, adding 5 per cent of sulfur to give it that tang, and small quantities of other elements to taste. (Use 'Mother Stellar's Elementary All-Spice' for best results.)

"Heat in a radioactive furnace until the mass is thoroughly melted and two mutually insoluble layers separate. (CAUTION. Do not heat too long, as prolonged heating will induce a desiccation that is not desirable.)

"Cool slowly till the crust hardens and a thin film of adhering gas and moisture appears. (If it does not appear, you have overheated.) Place in an orbit at a comfortable distance from a star and set to spinning. Then wait. In several billion years it will ferment at the surface. The fermented portion, called life, is considered the best part by connoisseurs."

15 The Trojan Hearse

The very first story I ever had published (never mind how long ago that was) concerned a spaceship that had come to grief in the asteroid zone. In it, I had a character comment on the foolhardiness of the captain in not moving out of the plane of the ecliptic (i.e. the plane of the earth's orbit, which is close to that in which virtually all the components of the solar system move) in order to go over or under the zone and avoid almost certain collision.

The picture I had in mind at that time was of an asteroidal zone as thickly strewn with asteroids as a beach was with pebbles. This is the same picture that exists, I believe, in the mind of almost all science-fiction writers and readers. Individual miners, one imagines, can easily hop from one piece of rubble to the next in search of valuable minerals. Vacationers can pitch their tents on one world and wave at the vacationers on neighboring worlds. And so on.

How true is this picture? The number of asteroids so far discovered is somewhat less than 2,000, but, of course, the actual number is far higher. I have seen estimates that place the total number at 100,000.

Most of the asteroids are to be found between the orbits of Mars and Jupiter and within 30° of the plane of the ecliptic. Now, the total volume of space between those orbits and within that tilt to the ecliptic is (let's see now—mumble, mumble, mumble) something like 200,000,000,-000,000,000,000,000,000 cubic miles. If we allow for a total quantity of 200,000 asteroids, to be on the safe side, then there is one asteroid for every 1,000,000,000,000-000,000,000 cubic miles.

This means that the average distance between asteroids

189

is 10,000,000 miles. Perhaps we can cut that down to 1,000,000 miles for the more densely populated regions. Considering that the size of most asteroids is under a mile in diameter, you can see that from any one asteroid you will in all probability see no others with the naked eye. The vacationer will be lonely and the miner will have a heck of a problem reaching the next bit of rubble.

In fact, astronauts of the future will in all probability routinely pass through the asteroid zone on their way to the outer planets and never see a thing. Far from being a dreaded sign of danger, the occasional cry of "asteroid in view" should bring all the tourists rushing to the portholes.

Actually, we mustn't think of the asteroidal zone as evenly strewn with asteroids. There are such things as clusters and there are also bands within the zone that are virtually empty of matter.

The responsibility for both situations rests with the planet Jupiter and its strong pull on the other components of the solar system.

As an asteroid in the course of its motion makes its closest approach to Jupiter (in the course of *its* motion), the pull of Jupiter on that asteroid reaches a maximum. Under this maximum pull, the extent by which an asteroid is pulled out of its normal orbit (is "perturbed") is also at a maximum.

Under ordinary circumstances, however, this approach of the asteroid to Jupiter occurs at different points in their orbits. Because of the rather elliptical and tilted orbits of most asteroids, the closest approach therefore takes place at varying angles, so that sometimes the asteroid is pulled forward at the time of its closest approach and sometimes backward, sometimes downward and sometimes upward. The net result is that the effect of the perturbations cancels out and that, in the long run, the asteroids will move in orbits that oscillate about some permanent average-orbit.

But suppose an asteroid circled about the sun at a mean distance of about 300,000,000 miles? It would then have a period of revolution of about six years as compared with Jupiter's period of twelve years.

If the asteroid were close to Jupiter at a given moment of time; then twelve years later, Jupiter would have made just one circuit and the asteroid just two circuits. Both would occupy the same relative positions again. This would repeat every twelve years. Every other revolution, the asteroid would find itself yanked in the same direction. The perturbations, instead of canceling out, would build up. If the asteroid was constantly pulled forward at its close approach, it would gradually be moved into an orbit slightly more distant from the sun, and its year would become longer. Its period of revolution would then no longer match Jupiter's, and the perturbations would cease building up.

If, on the other hand, the asteroid were pulled backward each time, it would gradually be forced into an orbit that was closer to the sun. Its year would become shorter; it would no longer match Jupiter's; and again the perturbations would cease building up.

The general effect is that no asteroid is left in that portion of the zone where the period of revolution is just half that of Jupiter. Any asteroid originally in that portion of the zone moves either outward or inward; it does not stay put.

The same is true of that region of the zone in which an asteroid would have a period of revolution equal to four years, for then it would repeat its position with respect to Jupiter every three revolutions. If it had a period of revolution equal to 4.8 years, it would repeat its position with respect to Jupiter every five revolutions, and so on.

The regions of the asteroid zone which have thus been swept clear of asteroids by Jupiter are known as "Kirkwood's gaps." They are so named because the American astronomer Daniel Kirkwood called attention to these gaps in 1876 and explained them properly.

An exactly analogous situation is also to be found in the case of Saturn's rings—which, is in fact, why we speak of "rings" rather than "ring."

The rings were first discovered by the Dutch astronomer Christian Huyghens in 1655. To him it seemed a simple ring of light circling Saturn but touching it nowhere. In 1675, however, the Italian-born French astronomer Giovanni Domenico Cassini noticed that a dark gap divided the ring into a thick and bright inner portion and a thinner and somewhat less bright outer portion. This gap, which is 3,000 miles wide, has been called the "Cassini division" ever since.

In 1850, a third, quite dim ring, closer to Saturn than are the others, was spied by the American astronomer George Phillips Bond. It is called the "crape ring" because it is so dim. The crape ring is separated from the inner bright ring by a gap of 1,000 miles.

In 1859, the Scottish mathematician Clerk Maxwell showed that from gravitational considerations, the ring could not be one piece of material but had to consist of numerous light-reflecting fragments that seemed one piece only because of their distance. The fragments of the crape ring are much more sparsely distributed than those of the bright rings, which is why the crape ring seems so dim. This theoretical prediction was verified when the period of revolution of the rings was measured spectroscopically and found to vary from point to point. After all, if the rings were one piece, the period or revolution would be everywhere the same.

The innermost portion of the crape ring is a mere 7,000 miles above Saturn's surface. Those particles move most rapidly and have the shortest distance to cover. They revolve about Saturn in about 3¼ hours.

As one moves outward in the rings, the particles move more slowly and must cover greater distances, which means that the period of revolution mounts. Particles at the outermost edge of the rings have a period or revolution of about 13½ hours.

If particles were to be found in Cassini's division, they would circle Saturn in a period of a little over 11 hours. But particles are not found in that region of the rings, which is why it stands out black against the brightness on either side.

Why?

Well, outside the ring system, Saturn possesses a family of nine satellites, each of which has a gravitational field that produces perturbation in the motion of the particles of the rings. Saturn's innermost satellite, Mimas, which lies only 35,000 miles beyond the outer edge of the rings, has a period of revolution of 22½ hours. Enceladus, the second satellite, has a period of 33 hours; and Tethys, the third satellite, a period of 44 hours.

Any particles in Cassini's division would have a period of revolution half that of Mimas, a third that of Enceladus, and a fourth that of Tethys. No wonder the region is swept clean. (Actually, the satellites are small bodies and their perturbing effect is insignificant on anything larger than the gravel that makes up the rings. If this were not so, the satellites themselves would by now have been forced out of their own too-closely matching orbits.)

As for the gap between the crape ring and the inner bright one, particles within it would circle Saturn in a little over seven hours, one-third the period of revolution of Mimas and one-sixth that of Tethys. There are other smaller divisions in the ring system which can be explained in similar fashion.

I must stop here and point out a curiosity that I have never seen mentioned. Books on astronomy always point out that Phobos, the inner satellite of Mars, revolves about Mars in less time than it takes Mars to rotate about its axis. Mars' period of rotation is 24½ hours while Phobos' period of revolution is only 7½ hours. The books then go on to say that Phobos is the only satellite in the system of which this is true.

Well, that is correct if we consider natural satellites of appreciable size. However, each particle in Saturn's rings is really a satellite, and if they are counted in, the situation changes. The period of rotation of Saturn about its axis is 10½ hours, and every particle in the crape ring and in the inner bright ring revolves about Saturn in less time than that. Therefore, far from there being only one satellite of the Phobos type, there are uncounted millions of them.

In addition, almost every artificial satellite sent up by the United States and the Soviet Union revolves about the earth in less than twenty-four hours. They, too, are of the Phobos type.

Gravitational perturbations act not only to sweep regions clear of particles but also to collect them. The most remarkable case is one where particles are collected not in a zone, but actually in a point.

To explain that, I will have to begin at the beginning. Newton's law of universal gravitation was a complete solution of the "two-body problem" (at least in classical physics, where the modern innovations of relativity and quantum theory are ignored). That is, if the universe contains only two bodies, and the position and motion of each are known, then the law of gravitation is sufficient to predict the exact relative positions of the two bodies through all of time, past and future.

However, the universe doesn't contain only two bodies. It contains uncounted trillions. The next step then is to build up toward taking them all into account by solving the "three-body problem." Given three bodies in the universe, with known position and motion, what will their relative positions be at all given times?

And right there, astronomers are stymied. There is no general solution to such a problem. For that reason there is no use going on to the "octillion-body problem" represented by the actual universe.

Fortunately, astronomers are not halted in a practical sense. The theory may be lacking but they can get along. Suppose, for instance, that it was necessary to calculate the orbit of the earth about the sun so that the relative positions could be calculated for the next million years. If the sun and the earth were all that existed, the problem would be a trivial one to solve. But the gravity of the moon must be considered, and of Mars and the other planets, and, for complete exactness, even the stars.

Fortunately, the sun is so much bigger than any other body in the vicinity and so much closer than any other really massive body, that its gravitational field drowns out all others. The orbit obtained for the earth by calculating a simple two-body situation is almost right. You then calculate the minor effect of the closer bodies and make corrections. The closer you want to pinpoint the exact orbit, the more corrections you must make, covering smaller and smaller perturbations.

The principle is clear but the practice can become tedious, to be sure. The equation that gives the motion of the moon with reasonable exactness covers many hundreds of pages. But that is good enough to predict the time and position of eclipses with great correctness for long periods of time into the future.

Nevertheless, astronomers are not satisfied. It is all very well to work out orbits on the basis of successive approximations, but how beautiful and elegant it would be to prepare an equation that would interrelate all bodies in a simple and grand way. Or three bodies, anyway.

The man who most closely approached this ideal was the French astronomer Joseph Louis Lagrange. In 1722, he actually found certain very specialized cases in which the three-body problem could be solved.

Imagine two bodies in space with the mass of body A at least 25.8 times that of body B, so that B can be said to revolve about a virtually motionless A, as Jupiter, for instance, revolves about the sun. Next imagine a third body, C, of comparatively insignificant mass, so that it does not disturb the gravitational relationship of A and B. Lagrange found that it was possible to place body C at certain points in relationship to bodies A and B, so that C could revolve about A in perfect step with B. In that way the relative positions of all three bodies would be known for all times.

There are five points at which body C can be placed; and they are, naturally enough, called "Lagrangian points." Three of them, L_1, L_2 and L_3, are on the line connecting A and B. The first point, L_1, places small body C between A and B. Both L_2 and L_3 lie on the line also, but on the other side of A in the first case and on the other side of B in the next.

These three Lagrangian points are not important. If any body located at one of those points, moves ever so slightly off position due to the perturbation of some body outside the system, the resulting effect of the gravitational fields of A and B is to throw C still farther off the point. It is like a long stick balanced on edge. Once that stick tips ever so slightly, it tips more and more and falls.

However, the final Lagrangian points are not on the line connecting bodies A and B. Instead, they form equilateral triangles with A and B. As B revolves about A, L_4 is the point that moves before B at constant angle of 60°, while L_5 moves behind it at a constant 60°.

These last two points are stable. If an object at either point moves slightly off position, through outside perturbations, the effect of the gravitational fields of A and B is to bring them back. In this way, objects at L_4 and L_5 oscillate about the true Lagrangian point, like a long stick balanced at the end of a finger which adjusts its position constantly to prevent falling.

Of course, if the stick tips *too* far out of vertical it will fall despite the balancing efforts of the finger. And if a body moves *too* far away from the Lagrangian point it will be lost.

At the time Lagrange worked this out, no examples of objects located at Lagrangian points were known anywhere in the universe. However, in 1906, a German astronomer, Max Wolf, discovered an asteroid, which he named Achilles after the Greek hero of the *Iliad*. It was unusually far out for an asteroid. In fact, it was as far from the sun as Jupiter was.

An analysis of its orbit showed that it always remained near the Lagrangian point, L_4, of the sun-Jupiter system. Thus, it stayed a fairly constant 480,000,000 miles ahead of Jupiter in its motion about the sun.

Some years later, another asteroid was discovered in the L_5 position of the sun-Jupiter system, and was named Patroclus, after Achilles' beloved friend. It moves about the sun in a position that is fairly constant 480,000,000 miles behind Jupiter.

Other asteroids were in time located at both points; at the present time, fifteen of these asteroids are known: ten in L_4 and five in L_5. Following the precedent of Achilles, all have been named for characters in the *Iliad*. And since the *Iliad* deals with the Trojan War, all the bodies in both positions are lumped together as the "Trojan asteroids."

Since the asteroids at position L_4 include Agamemnon, the Greek leader, they are sometimes distinguished as the "Greek group." The asteroids at position L_5 include the one named for the Trojan king Priamus (usually known as

"Priam" in English versions of the *Iliad*), and are referred to as the "pure Trojan group."

It would be neat and tidy if the Greek group contained only Greeks and the pure Trojan group only Trojans. Unfortunately, this was not thought of. The result is that the Trojan hero Hector is part of the Greek group and the Greek hero Patroclus is part of the pure Trojan group. It is a situation that would strike any classicist with apoplexy. It makes even myself feel a little uncomfortable, and I am only the very mildest of classicists indeed.

The Trojan asteroids remain the only known examples of objects at Lagrangian points. They are so well known, however, that L_4 and L_5 are commonly known as "Trojan positions."

External perturbing forces, particularly that of the planet Saturn, keep the asteroids oscillating about the central points. Sometimes the oscillations are wide; a particular asteroid may be as much as 100,000,000 miles from the Lagrangian point.

Eventually, a particular asteroid may be pulled too far outward, and would then adopt a non-Trojan orbit. On the other hand, some asteroid now independent, may happen to be perturbed into a spot close to the Lagrangian points and be trapped. In the long run, the Trojan asteroids may change identities, but there will always be some there.

Undoubtedly, there are many more than fifteen Trojan asteroids. Their distance from us is so great that only fairly large asteroids, close to one hundred miles in diameter, can be seen. Still, there are certainly dozens and even hundreds of small chunks, invisible to us, that chase Jupiter or are chased in an eternal race that nobody wins.

There must be many Trojan situations in the universe. I wouldn't be surprised if every pair of associated bodies which met the 25.8 to 1 mass-ratio requirement was accompanied by rubble of some sort at the Trojan positions.

Knowing that the rubble exists doesn't mean that it can be spotted, however; certainly it can be detected nowhere outside the solar system. Three related stars could be spotted, of course, but for a true Trojan situation, one body must be of insignificant mass, and it could not be seen by any technique now at our disposal.

Within the solar system, by far the largest pair of bodies are the sun and Jupiter. The bodies strapped at the Lagrangian points of that system could be fairly large and yet remain negligible in mass in comparison to Jupiter.

The situation with respect to Saturn would be far less favorable. Since Saturn is smaller than Jupiter, the asteroids at the Trojan position associated with Saturn would be smaller on the average. They would be twice as far from us as those of Jupiter are, so that they would also be dimmer. They would thus be very difficult to see; and the fact of the matter is that no Saturnian Trojans have been found. The case is even worse for Uranus, Neptune and Pluto.

As for the small inner planets, there any rubble in the Trojan position must consist of small objects indeed. That alone would make them nearly impossible to see, even if they existed. In addition, particularly in the case of Venus and Mercury, they would be lost in the glare of the sun.

In fact, astronomers do not really expect to find the equivalent of Trojan asteroids for any planet of the solar system other than Jupiter, until such time as an astronomical laboratory is set up outside the earth or, better yet, until spaceships actually explore the various Lagrangian points.

Yet there is one exception to this, one place where observation from the earth's surface can turn up something and, in fact, may have done so. That is a Lagrangian point that is not associated with a sun-planet system, but with a planet-satellite system. Undoubtedly you are ahead of me and know that I am referring to the earth and the moon.

The fact that the earth has a single satellite was known as soon as man grew intelligent enough to become a purposeful observer. Modern man with all his instruments has never been able to find a second one. Not a natural one, anyway. In fact, astronomers are quite certain that, other than the moon itself, no body that is more than, say, half a mile in diameter, revolves about the earth.

This does not preclude the presence of any number of very small particles. Data brought back by artificial satellites would seem to indicate that the earth is surrounded by a ring of dust particles something after the fashion of Saturn, though on a much more tenuous scale.

Visual observation could not detect such a ring except in places where the particles might be concentrated in unusually high densities. The only spots where the concentration could be great enough would be at the Lagrangian points, L_4 and L_5, of the earth-moon system. (Since the earth is more than 25.8 times as massive as the moon— it is 81 times as massive in point of fact—objects at those points would occupy a stable position.)

Sure enough, in 1961, a Polish astronomer, K. Kordylewski, reported actually spotting two very faintly luminous patches in those positions. Presumably, they represent dust clouds trapped there.

And in connection with these "cloud satellites," I have thought up a practical application of Lagrangian points which, as far as I know, is original.

As we all know, one of the great problems brought upon us by the technology of the space age is that of the disposal of radioactive waters. Many solutions have been tried or have been suggested. The wastes are sealed in strong containers or, as is suggested, fused in glass. They may be buried underground, stored in salt-mines or dropped into an abyss.

No solution that leaves the radioactivity upon the earth, however, is wholly satisfactory; so some bold souls have suggested that eventually measures will be taken to fire the wastes into space.

The safest procedure one can possibly imagine is to shoot these wastes into the sun. This, however, is not an easy thing to do at all. It would take less energy to shoot them to the moon, but I'm sure that astronomers would veto that. It would be still easier simply to shoot them into an orbit about the sun, and easiest of all to shoot them into an orbit about the earth.

In either of these latter cases, however, we run the risk, in the long run, of cluttering up the inner portions of the solar system, particularly the neighborhood of the earth, with gobs of radioactive material. We would be living in the midst of our own refuse, so to speak.

Granted that space is large and the amount of refuse, in comparison, is small, so that collisions or near-collisions between spaceships and radioactive debris would be highly improbable, it could still lead to trouble in the long run.

Consider the analogy of our atmosphere. All through history, man has freely poured gaseous wastes and smoky particles into it in the certainty that all would be diluted far past harm; yet air pollution has now become a major problem. Well, let's not pollute space.

One way out is to concentrate our wastes into one small portion of space and make sure it stays there. Those regions of space can then be marked off-limits and everything else will be free of trouble.

To do this, one would have to fire the wastes to one or the other of the Trojan positions associated with the earth-moon system in such a way as to leave it trapped there. Properly done, the waste would remain at those points, a quarter-million miles from the moon and a quarter-million miles from the earth, for indefinite periods, certainly long enough for the radiation to die down to nondangerous levels.

Naturally, the areas would be a death trap for any ship passing through—a kind of "Trojan hearse," in fact. Still, it would be a small price to pay for solving the radioactive ash disposal problem, just as this pun is a small price to pay for giving me a title for the chapter.

16 By Jove!

Suppose we ask ourselves a question: On what world of the solar system (other than earth itself, of course) are we most likely to discover life?

I imagine I can plainly hear the unanimous answering shout, *"Mars!"*

The argument goes, and I know it by heart, because I have used it myself a number of times, that Mars may be a little small and a little cold and a little short on air, but it isn't too small, too cold, or too airless to support the equivalent of primitive plant life. On the other hand, Venus and Mercury are definitely too hot, the moon is airless, and the remaining satellites of the solar system, and the planetoids as well (to say nothing of Pluto), are too cold, too small, or both.

And then we include a phrase which may go like this: "As for Jupiter, Saturn, Uranus and Neptune, we can leave them out of consideration altogether."

However, Carl Sagan, an astronomer at Harvard University, doesn't take that attitude at all, and a recent paper of his on the subject* has lured me into doing a bit of thinking on the subject of the outer planets.

Before Galileo's time, there was nothing to distinguish Jupiter and Saturn (Uranus and Neptune not having yet been discovered) from the other planets, except for the fact that they moved more slowly against the starry background than did the other planets and were, therefore, presumably farther from the earth.

---

*Professor Sagan was kind enough to send me a reprint of the paper, knowing that I would be interested in the subject. I am very grateful to him and to some others who have been equally kind in the matter of calling their work to my attention.

The telescope, however, showed Jupiter and Saturn as discs with angular widths that could be measured. When the distances of the planets were determined, those angular widths could be converted into miles, and the result was a shocker. As compared with an earthly equatorial diameter of 7,950 miles, Jupiter's diameter across its equator was 88,800, while Saturn's was 75,100.

The outer planets were giants!

The discovery of Uranus in 1781 and Neptune in 1846 added two more not-quite-so-giants, for the equatorial diameter of Uranus is 31,000 miles and that of Neptune, at latest measurement, is about 28,000 miles.

The disparity in size between these planets and our own tight little world is even greater if one considers volume, because that varies as the cube of the diameter. In other words, if the diameter of Body A is ten times the diameter of Body B, then the volume of Body A is ten times ten times ten, or a thousand times the volume of Body B. Thus, if we set the volume of the earth equal to 1, here are the volumes of the giants:

Jupiter	—	1,300
Saturn	—	750
Uranus	—	60
Neptune	—	40

Each of the giants has satellites. It is easy to determine the distance of the various satellites from the center of the primary planet by measuring the angular separation. It is also easy to time the period of revolution of the satellite. From those two pieces of datum, one can quickly obtain the mass of the primary. (It is because Venus and Mercury have no satellites that we are less certain about their mass than we are, for instance, about Neptune's.)

In terms of mass, the giants remain giants, naturally. If the mass of earth is taken as 1, the masses of the giants are:

Jupiter	—	318
Saturn	—	95
Uranus	—	15
Neptune	—	17

The four giants contain virtually all the planetary mass of the solar system, Jupiter alone possessing about 70 per cent of the total. The remaining planets, plus all the satellites, planetoids, comets and, for that matter, meteoroids, contain well under 1 per cent of the total planetary mass, Outside intelligences, exploring the solar system with true impartiality, would be quite likely to enter the sun in their records thus: Star X, spectral class GO, 4 planets plus debris.

But take another look at the figures on mass. Compare them with those on volume and you will see that the mass is consistently low. In other words, Jupiter takes up 1,300 times as much room as the earth does, but contains only 318 times as much matter. The matter in Jupiter must therefore be spread out more loosely, which means, in more formal language, that Jupiter's density is less than that of the earth.

. If we set the earth's density equal to 1, then we can obtain the densities of the giants by just dividing the figure for the relative mass by the figure of the relative volume. The densities of the giants are:

Jupiter	—	0.280
Saturn	—	0.125
Uranus	—	0.250
Neptune	—	0.425

On this same scale of densities, the density of water is 0.182. As you see, then, Neptune, the densest of the giants, is only about 2¼ times as dense as water, while Jupiter and Uranus are only 1½ times as dense, and Saturn is actually less dense than water.

I remember seeing an astronomy book that dramatized this last fact by stating that if one could find an ocean large enough, Saturn would float in it, less than three-fourths submerged. And there was a very impressive illustration, showing Saturn, rings and all, floating in a choppy sea.

But don't ministerpret this matter of density. The first thought anyone naturally might have is that because Saturn's overall density is less than that of water, it must be made of some corklike material. This, however, is not so, as I can explain easily.

Jupiter has a striped or banded appearance, and certain features upon its visible surface move around the planet at a steady rate. By following those features, the period of rotation can be determined with a high degree of precision; it turns out to be 9 hours, 50 minutes, and 30 seconds. (With increasing difficulty, the period of rotation can be determined for the more distant giants as well.)

But here a surprising fact is to be noted. The period of rotation I have given is that of Jupiter's equatorial surface. Other portions of the surface rotate a bit more slowly. In fact, Jupiter's period of rotation increases steadily as the poles are approached. This alone indicates we are not looking at a solid surface, for that would have to rotate all in one piece.

The conclusion is quite clear. What we see as the surface of Jupiter, and of the other giants, are the clouds of its atmosphere. Beneath those clouds must be a great depth of atmosphere, far thicker than our own, and yet far less dense than rock and metal. It is because the atmosphere of the giant planets is counted in with their volume that their density appears so low. If we took into account only the core of the planet, underlying the atmosphere, we would find a density as great as that of earth's, or, most likely, greater.

But how deep is the atmosphere?

Consider that, fundamentally, the giant planets differ from the earth chiefly in that, being further from the sun and therefore colder through their history, they retain a much larger quantity of the light elements—hydrogen, helium, carbon, nitrogen and oxygen. Helium forms no compounds but remains as a gas. Hydrogen is present in large excess so it remains as a gas, too, but it also forms compounds with carbon, nitrogen and oxygen, to form methane, ammonia and water, respectively. Methane is a gas, and, at the earth's temperature, so is ammonia, but water is a liquid. If the earth's temperature were to drop to —100° C. or below, both ammonia and water would be solid, but methane would still be a gas.

As a matter of fact, all this is not merely guesswork. Spectroscopic evidence does indeed show that Jupiter's atmosphere is hydrogen and helium in a three-to-one ratio with liberal admixtures of ammonia and methane. (Water is not detected, but that may be assumed to be frozen out.)

Now, the structure of the earth can be portrayed as a central solid body of rock and metal (the lithosphere), surrounded by a layer of water (the hydrosphere), which is in turn surrounded by a layer of gas (the atmosphere).

The light elements in which the giant planets are particularly rich would add to the atmosphere and to the hydrosphere, but not so much to the lithosphere. The picture would therefore be of a central lithosphere larger than that of the earth, but not necessarily enormously larger, surrounded by a gigantic hydrosphere and an equally gigantic atmosphere.

But how gigantic is gigantic?

Here we can take into consideration the polar flattening of the giants. Thus, although Jupiter is 88,800 miles in diameter along the equator, it is only 82,800 miles in diameter from pole to pole. This is a flattening of 7 per cent, compared to a flattening of about .33 per cent for the earth. Jupiter has a visibly elliptical appearance for that reason. Saturn's aspect is even more extreme, for its equatorial diameter is 75,100 miles while its polar diameter is 66,200 miles, a flattening of nearly 12 per cent. (Uranus and Neptune are less flattened than are the two larger giants.)

The amount of flattening depends partly on the velocity of rotation and the centrifugal effect which is set up. Jupiter and Saturn, although far larger than the earth, have periods of rotation of about 10 hours as compared with our own 24. Thus, the Jovin surface, at its equator, is moving at a rate of 25,000 miles per hour, while the earth's equatorial surface moves only at a rate of 1,000 miles an hour. Naturally, Jupiter's surface is thrown farther outward than earth's is (even against Jupiter's greater gravity), so that the giant planet bulges more at the equator and is more flattened at the poles.

However, Saturn is distinctly smaller than Jupiter and has a period of rotation some twenty minutes longer than that of Jupiter. It exerts a smaller centrifugal effect at the equator; and even allowing for its smaller gravity, it should be less flattened at the poles than Jupiter is. However, it is more flattened. The reason for this is that the degree of flattening depends also on the distribution of density, and if Saturn's atmosphere is markedly thicker than Jupiter's flattening will be greater.

The astonomer Rupert Wildt has estimated what the size of the lithosphere, hydrosphere and atmosphere would have to be on each planet in order to give it the overall density it was observed to have plus its polar flattening. (This picture is not accepted by astronomers generally, but let's work with it anyway.) The figures I have seen are included in the following table, to which I add figures for the earth as a comparison:

	Lithosphere (radius in miles)	Hydrosphere (thickness in miles)	Atmosphere (thickness in miles)
Jupiter	18,500	17,000	8,000
Saturn	14,000	8,000	16,000
Uranus	7,000	6,000	3,000
Neptune	6,000	6,000	2,000
Earth	3,975	2	8*

As you see, Saturn, though smaller than Jupiter, is pictured as having a much thicker atmosphere, which accounts for its low overall density and its unusual degree of flattening. Neptune has the shallowest atmosphere and is therefore the densest of the giant planets.

Furthermore, you can see that the earth isn't *too* pygmyish in comparison with the giants, if the lithosphere alone is considered. If we assume that the lithospheres are

*I know that the atmosphere is thicker than eight miles, and that in fact it has no fixed thickness. However, I am taking the earth's atmosphere—and shall later calculate its volume—only to the top of its cloud layers, which is what we do for the giant planets.

all of equal density and set the mass of earth's lithosphere equal to 1, then the masses of the others are:

Jupiter	—	100
Saturn	—	45
Uranus	—	5½
Neptune	—	3½

It is the disparity of the hydrosphere and atmosphere that blows up the giants to so large a size.

To emphasize this last fact, it would be better to give the size of the various components in terms of volume rather than of thickness. In the following table the volumes are therefore given in trillions of cubic miles. Once again, the earth is included for purposes of comparison:

	Lithosphere volume	*Hydrosphere volume*	*Atmosphere volume*
Jupiter	27	161	155
Saturn	11.5	33	185
Uranus	1.4	7.8	8.4
Neptune	0.9	6.4	4.2
Earth	0.26	0.00033	0.0011

As you can see at a glance, the lithosphere of the giant planets makes up only a small part of the total volume, whereas it makes up almost all the volume of the earth. This shows up more plainly if we set up the volume of each component as a percentage of its planet's total volume. Thus:

	Lithosphere (% of planet's volume)	*Hydrosphere (% of planet's volume)*	*Atmosphere (% of planet's volume)*
Jupiter	7.7	47.0	45.3
Saturn	4.8	14.4	80.8
Uranus	8.0	44.3	47.7
Neptune	8.0	55.5	36.5
Earth	99.45	0.125	0.425

The difference can't be made plainer. Whereas the earth is about 99.5 per cent lithosphere, the giant planets are only 8 per cent, or less, lithosphere. About one-third of Neptune's apparent volume is gas. In the case of Jupiter and Uranus, the gas volume is one-half the total, and in the case of Saturn, the least dense of the four, the gas volume is fully four-fifths of the total. The giant planets are sometimes called the "gas giants" and, as you see, that is a good name, particularly for Saturn.

This is a completely alien picture we have drawn of the giant planets. The atmospheres are violently poisonous, extremely deep and completely opaque, so that the surface of the planet is entirely and permanently dark even on the "sunlit side." The atmospheric pressure is gigantic; and from what we can see of the planets, the atmosphere seems to be beaten into the turmoil of huge storms.

The temperatures of the planets are usually estimated as ranging from a −100° C. maximum for Jupiter to a −230° C. minimum for Neptune, so that even if we could survive the buffeting and the pressures and the poisons of the atmosphere, we would land on a gigantic, planet-covering, thousands-of-miles-thick layer of ammoniated ice.

Not only is it inconceivable for man to land and live on such a planet, but it seems inconceivable that any life at all that even remotely resembles our own could live there.

Are there any loopholes in this picture?

Yes, a very big one, possibly, and that is the question of the temperature. Jupiter may not be nearly as cold as we have thought.

To be sure, it is about five times as far from the sun as we are, so that it receives only one twenty-fifth as much solar radiation. However, the crucial point is not how much radiation it receives but how much it keeps. Of the light it receives from the sun, four-ninths is reflected and the remaining five-ninths is absorbed. The absorbed portion does not penetrate to the planetary surface as light, but it gets there just the same—as heat.

The planet would ordinarily radiate this heat as long-wave infrared, but the components of Jupiter's atmosphere, notably the ammonia and methane, are quite opaque to infrared, which is therefore retained forcing the temperature to rise. It is only when the temperature is quite high that enough infrared can force its way out of the atmosphere to establish a temperature equilibrium.

It is even possible that the surface temperature of Jupiter, thanks to this "greenhouse effect," is as high as that of earth. This is not a matter of theory only, for the radio-wave emission by Jupiter, discovered first in 1955, seems to indicate an atmospheric temperature considerably higher than that which had long been considered likely.

The other giant planets may also have temperatures higher than those usually estimated, but the final equilibrium would very likely be lower than that of Jupiter's, since the other planets are further from the sun. Perhaps Jupiter is the only giant planet with a surface temperature above 0° C.

This means that Jupiter, of all the giant planets, would be the only one with a liquid hydrosphere. Jupiter would have a vast ocean, covering the entire planet (by the Wildt scheme) and 17,000 miles deep.

On the other hand, Venus also has an atmosphere that exerts a greenhouse effect, raising its surface temperature to a higher level than had been supposed. Radio-wave emission from Venus indicates its surface temperature to be much higher than the boiling point of water, so that the surface of Venus is powdery dry with all its water supply in the cloud layer overhead.

A strange picture. The planetary ocean that has been so time-honored a science-fictional picture of Venus has been pinned to the wrong planet all along. It is Jupiter that has the world-wide ocean, by Jove!

Considering the Jovian ocean, Professor Sagan (to whom I referred at the beginning of this chapter) says: "At the present writing, the possibility of life on Jupiter seems somewhat better than the possibility of life on Venus."

This is a commendably cautious statement, and as far as a scientist can be expected to go in a learned journal.

However, I, myself, on this particular soapbox, don't have to be cautious at all, so I can afford to be much more sanguine about the Jovian ocean. Let's consider it for a moment.

If we accept Wildt's picture, it is a big ocean, nearly 500,000 times as large as earth's ocean and, in fact, 620 times as voluminous as all the earth. This ocean is under the same type of atmosphere that, according to current belief, surrounded the earth at the time life developed on our planet. All the sample compounds—methane, ammonia, water, dissolved salts—would be present in unbelievable plenty of earthly standards.

Some source of energy is required for the building up of these organic compounds, and the most obvious one is the ultraviolet radiation of the sun. The quantity of ultraviolet rays that reaches Jupiter is, as aforesaid, only one twenty-fifth that which reaches the earth, and none of it can get very far into the thick atmosphere.

Nevertheless, the ultraviolet rays must have some effect, because the colored bands in the Jovian atmosphere are very likely to consist of free radicals (that is, energetic molecular fragments) produced out of ordinary molecules by the ultraviolet.

The constant writhing of the atmosphere would carry the free radicals downward where they could transfer their energy by reacting with simple molecules to build up complex ones.

Even if ultraviolet light is discounted as an energy source, two other sources remain. There is first, lightning. Lightning in the thick soup that is called a Jovian atmosphere may be far more energetic and continuous than it ever is or was on earth. Secondly, there is always natural radioactivity.

Well, then, why can't the Jovian ocean breed life? The temperature is right. The raw material is there. The energy supply is present. All the requirements that were sufficient to produce life in earth's primordial ocean are present also on Jupiter (if the picture drawn in this article is correct), only more and better.

One might wonder whether life could withstand the Jovian atmospheric pressures and storms, to say nothing of the Jovian gravity. But the storms, however violent, could only roil up the outer skin of a 17,000-mile-deep ocean. A few hundred feet below the surface, or a mile below, if you like, there would be nothing but the slow ocean currents.

As for gravity, forget it. Life within the ocean can ignore gravity altogether, for buoyancy neutralizes its effects, or almost neutralizes it.

No, none of the objections stand up. To be sure, life must originate and develop in the absence of gaseous oxygen, but that is exactly one of the conditions under which life originated and developed on earth. There are living creatures on earth right now that can live without oxygen.

So once again let's ask the question: On what world of the solar system (other than earth itself, of course) are we most likely to discover life?

And now, it seems to me, the answer must be: On Jupiter, by Jove!

Of course, life on Jupiter would be pitifully isolated. It would have a vast ocean to live in, but the far, far vaster outside universe would be closed forever to them.

Even if some forms of Jovian life developed an intelligence comparable to our own (and there are reasonable arguments to suggest that true sea life—and before you bring up the point, dolphins are descendants of land-living creatures—would not develop such intelligence), they could do nothing to break the isolation.

It is highly unlikely that even a manlike intelligence could devise methods that would carry itself out of the ocean, through thousands of miles of violent, souplike atmosphere, against Jupiter's colossal gravity, in order to reach Jupiter's inmost satellite and, from its alien surface, observe the universe.

And as long as life remained in the Jovian ocean, it would receive no indication of an outside universe, except for a non-directed flow of heat, and excessively feeble microwave radiation from the sun and a few other spots. Considering the lack of supporting information, the microwaves would be as indecipherable a phenomenon as one could imagine, even if it were sensed.

But let's not be sad; let's end on a cheerful note.

If the Jovian ocean is as rich in life as our own is, then $\frac{1}{70,000}$ of its mass would be living matter. In other words, the total mass of sea life on Jupiter would then be one-eighth the mass of our moon, and that's a lot of mass for a mess of fish.

What fishing-grounds Jupiter would make if it could be reached somehow!

And, in view of our population explosion, just one question to ponder over. . . . Do you suppose that Jovian life might be edible?

17 Superficially Speaking

For the last century, serious science-fiction writers, from Edgar Allan Poe onward, have been trying to reach the moon; and now governments are trying to get into the act. It kills some of the romance of the deal to have the project become a "space spectacular" designed to show up the other side, but if that's what it takes to get there, I suppose we can only sigh and push on.

So far, however, governments are only interested in *reaching* the moon, and as science-fiction fans we ought to remain one step ahead of them and keep our eyes firmly fixed on *populating* the moon. Naturally, we can ignore such little problems as the fact that air and water are missing on the moon. Perhaps we can bake water out of the deep-lying rock and figure out ways of chipping oxygen out of silicates. We can live underground to get away from the heat of the day and the cold of the night.

In fact, with the sun shining powerfully down from a cloudless sky for two weeks at a time, solar batteries might be able to supply moon colonists with tremendous quantities of energy.

Maybe the land with the high standard of living of the future will be up there in the sky. Etched into some of the craters, perhaps, large enough to be clearly seen through a small telescope, could be a message that starts: "Send me your tired, your poor, your huddled masses yearning to breathe free . . ."

Who knows?

But if the moon is ever to be a second earth and is to siphon off some of our population, there is a certain significant statistic about it that we ought to know. That is, its size.

The first question is, <u>what do we mean by "size"?</u>

The size of the moon is most often given in terms of its diameter, because once the moon's distance has been determined, its diameter can be obtained by direct measurement.

Since the moon's diameter is 2,160 miles and the earth's is 7,914 miles, most people cannot resist the temptation of saying that the moon is one-quarter the size of the earth, or that the earth is four times the size of the moon. (The exact figure is that the moon is 0.273 times the size of the earth, from this viewpoint, or that the earth is 3.66 times the size of the moon.)

All this makes the moon appear quite a respectably-sized world.

But, let's consider size from a different standpoint. Next to diameter, the most interesting statistic about a body of the solar system is its mass, for upon that depends the gravitational force it can exert.

Now, mass varies as the cube of the diameter, all things being equal. If the earth is 3.66 times the size of the moon, diameterwise, it is $3.66 \times 3.66 \times 3.66$ or 49 times the size of the moon, masswise. (Hmm, there's something to be said for this Madison Avenue speech monstrosity, conveniencewise.) But that is only if the densities of the two bodies being compared are the same.

As it happens, the earth is 1.67 times as dense as the moon, so that the discrepancy in mass is even greater than a simple cubing would indicate. Actually, the earth is 81 times as massive as the moon.

This is distressing because now, suddenly, the moon has grown a bit pygmyish on us, and the question arises as to which we ought really to say. Is the moon one-quarter the size of the earth or is it only $\frac{1}{81}$ the size of the earth?

Actually, we ought to use whichever comparison is meaningful under a particular set of circumstances, and as far as populating the moon is concerned, neither is directly meaningful. What counts is the surface area, the *superficial* size of the moon.

On any sizable world, under ordinary circumstances, human beings will live on the surface. Even if they dig underground to escape an unpleasant environment, they will do so only very slightly, when compared with the total diameter, on any world the size of the earth or even that of the moon.

Therefore, the question that ought to agitate us with respect to the size of the moon is: What is its surface area in comparison with that of the earth? In other words, what is its size, superficially speaking?

This is easy to calculate because surface area varies as the square of the diameter. Here density has no effect and need not be considered. If the earth has a diameter 3.66 times that of the moon, it has a surface area 3.66×3.66 or 13.45 times that of the moon.

But this doesn't satisfy me. The picture of a surface that is equal to $\frac{1}{13.45}$ that of the earth isn't dramatic enough. What does it mean exactly? Just how large is such a surface?

I've thought of an alternate way of dramatizing the moon's surface, and that of other areas, and it depends on the fact that a good many Americans these days have been jetting freely about the United States. This gives them a good conceptual feeling of what the area of the United States is like, and we can use that as a unit. The area of all fifty states is 3,628,000 square miles and we can call that 1 USA unit.

To see how this works, look at Table 1, which includes a sampling of geographic divisions of our planet with their areas given in USAs.

TABLE 1

Geographic division	Area (in USAs)
Australia	0.82
Brazil	0.91
Canada	0.95
United States	1.00
Europe	1.07
China	1.19
Arctic Ocean	1.50

Antarctica	1.65
South America	1.90
Soviet Union	2.32
North America	2.50
Africa	3.20
Asia	4.70
Indian Ocean	7.80
Atlantic Ocean	8.80
Total land surface	17.50
Pacific Ocean	17.60
Total water surface	36.80
Total surface	54.30

Now, you see, when I say that the moon's surface is 4.03 USAs, you know at once that the colonization of the moon will make available to humanity an area of land equal to four times that of the United States or 1.75 times that of the Soviet Union. To put it still another way, the area of the moon is just about halfway between that of Africa and Asia.

But let's go further and assume that mankind is going to colonize all the solar system that it can colonize or that is worth colonizing. When we say "can colonize," we eliminate, at least for the foreseeable future, the "gas giants," that is Jupiter, Saturn, Uranus and Neptune. (For some comments on them, however, see the previous chapter.)

That still leaves four planets: Mercury, Venus, Mars, and (just to be complete—and extreme) Pluto. In addition, there are a number of sizable satellites, aside from our own moon, that are large enough to seem worth colonizing. These include the four large satellites of Jupiter (Io, Europa, Ganymede and Callisto), the two large satellites of Saturn (Titan and Rhea), and Neptune's large satellite (Triton).

The surface areas of these bodies are easily calculated; the results are given in Table 2, with the earth and moon included for comparison. As you can see, if we exclude the sun and the gas giants, there are a round dozen bodies in the solar system with a surface area in excess of 1 USA, and a thirteenth with an area just short of that figure.

TABLE 2

Planet or satellite	Surface area (USAs)
Earth	54.3
Pluto	(54)??
Venus	49.6
Mars	15.4
Callisto	9.0
Ganymede	8.85
Mercury	8.30
Titan	7.30
Triton	6.80
Io	4.65
Moon	4.03
Europa	3.30
Rhea	0.86

The total surface area available on this baker's dozen of worlds is roughly equal to 225 USAs. Of this, the earth itself represents fully one quarter, and the earth is already colonized, so to speak, by mankind. Another quarter is represented by Pluto, the colonization of which, with the best will in the world, must be considered as rather far off.

Of what is left (about 118 USAs), Venus, Mars and the moon make up some five-ninths. Since these represent the worlds that are closest and therefore the most easily reached and colonized, there may be quite a pause before humanity dares the sun's neighborhood to reach Mercury, or sweeps outward to the large outer satellites. It might seem that the extra pickings are too slim.

However, there are other alternatives, as I shall explain.

So far, I have not considered objects of the solar system that are less than 1,000 miles in diameter (which is the diameter of Rhea). At first glance, these might be considered as falling under the heading of "not worth colonizing" simply because of the small quantity of surface area they might be expected to contribute. In addition, gravity would be so small as to give rise to physiological and technological difficulties, perhaps.

However, let's ignore the gravitational objection, and concentrate on the surface area instead.

Are we correct in assuming that the surface area of the minor bodies is small enough to ignore? There are, after all, twenty-three satellites in the solar system with diameters of less than 1,000 miles, and that's a respectable number. On the other hand, some of these satellites are quite small. Deimos, the smaller satellite of Mars, has a diameter that isn't more than 7.5 miles.

To handle the areas of smaller worlds, let's make use of another unit. The largest city of the United States, in terms of area, at least, is Los Angeles, which covers 450 square miles. We can set that equal to 1 LA. This is convenient because it means there are just about 8,000 LAs in 1 USA.

TABLE 3

Satellite (primary)	Surface area (LAs)
Iapetus (Saturn)	4,450
Tethys (Saturn)	3,400
Dione (Saturn)	3,400
Titania (Uranus)	2,500
Oberon (Uranus)	2,500
Mimas (Saturn)	630
Enceladus (Saturn)	630
Ariel (Uranus)	630
Umbriel (Uranus)	440
Hyperion (Saturn)	280
Phoebe (Saturn)	280
Nereid (Neptune)	120
Amalthea (Jupiter)	70
Miranda (Uranus)	45
VI (Jupiter)	35
VII (Jupiter)	6.5
VIII (Jupiter)	6.5
IX (Jupiter)	1.5
XI (Jupiter)	1.5
XII (Jupiter)	1.5
Phobos (Mars)	1.5
X (Jupiter)	0.7
Deimos (Mars)	0.4

A comparison of the surface areas of the minor satellites of the solar system is presented in Table 3. (I'll have to point out that the diameters of all these satellites are quite uncertain and that the surface areas as given are equally uncertain. However, they are based on the best information available to me.)

The total areas of the minor satellites of the solar system thus comes to just under 20,000 LAs or, dividing by 8,000, to about 2.5 USAs. All twenty-three worlds put together have little more than half the surface area of the moon, or, to put it another way, have just about the area of North America.

This would seem to confirm the notion that the minor satellites are not worth bothering about, but . . . let's think again. All these satellites, lumped together, have just a trifle over one-sixth the volume of the moon, and yet they have more than half its surface area.

This should remind us that the smaller a body, the larger its surface area in porportion to its volume. The surface area of any sphere is equal to $4\pi r^2$, where r is its radius. This means that the earth, with a radius of roughly 4,000 miles, has a surface area of roughly 200,000,000 square miles.

But suppose the material of the earth is used to make up a series of smaller worlds each with half the radius of the earth. Volume varies as the cube of the radius, so the material of the earth can make up no less than eight "half-earths," each with a radius of roughly 2,000 miles. The surface area of each "half-earth" would be roughly 50,000,-000 square miles, and the total surface area of all eight "half-earths" would be 400,000,000 square miles, or twice the area of the original earth.

If we consider a fixed volume of matter, then, the smaller the bodies into which it is broken up, the larger the total surface area it exposes.

You may feel this analysis accomplishes nothing, since the twenty-three minor satellites do not, in any case, have much area. Small though they are, the total area comes to that of North America and no more.

Ah, but we are not through. There are still the minor planets, or asteroids.

It is estimated that all the asteroids put together have a mass about 1 per cent that of the earth. If all of them were somehow combined into a single sphere, with an average density equal to that of the earth, the radius of that sphere would be 860 miles and the diameter, naturally, 1,720 miles. It would be almost the size of one of Jupiter's satellites, Europa, and its surface area would be 2.6 USAs, or just about equal to that of all the minor satellites put together.

But the asteroids do not exist as this single fictitious sphere but as a large number of smaller pieces, and here is where the increase in surface area comes in. The total number of asteroids is estimated to be as high as 100,000; and if that figure is correct, then the average asteroid has a diameter of 35 miles, and a total surface area of all 100,000 would then come to as much as 130 USAs.

This means that the total surface area of the asteroids is equal to slightly more than that of the earth, Venus, Mars, and the moon all lumped together. It is 7.5 times the area of the earth's land surface. Here is an unexpected bonanza.

Furthermore, we can go beyond that. Why restrict ourselves only to the surface of the worlds? Surely we can dig into them and make use of the interior materials otherwise beyond our reach. On large worlds, with their powerful gravitational forces, only the outermost skin can be penetrated, and the true interior seems far beyond our reach. On an asteroid, however, gravity is virtually nil and it would be comparatively easy to hollow it out altogether.

I made use of this notion in a story I once wrote which was set on a fictitious asteroid called Elsevere. A visitor from earth is being lectured by one of the natives, as follows:

"We are not a small world, Dr. Lamorak; you judge us by two-dimensional standards. The surface area of Elsevere is only three-quarters that of the state of New York, but that's irrelevant. Remember we can occupy, if we wish, the entire interior of Elsevere. A sphere of fifty miles' radius has a volume of well over half a million cubic miles. If all of Elsevere were occupied by levels fifty feet apart, the total surface area within the planetoid would be 56,000,-000 square miles, and that is equal to the total land area of earth. And none of these square miles, doctor, would be unproductive."

Well, that's for an asteroid 50 miles in radius and, consequently, 100 miles in diameter. An asteroid that is 35 miles in diameter would have only about $\frac{1}{27}$ the volume, and its levels would offer a surface area of only 2,000,000 square miles, which is nevertheless over half the total area of the United States (0.55 USAs, to be exact).

One small 35-mile-diameter asteroid would then offer as much living space as the moderately large Saturnian satellite Iapetus, if, in the latter case, only surface area were considered.

The material hollowed out of an asteroid would not be waste, either. It could be utilized as a source of metal, and of silicates. The only important elements missing would be hydrogen, carbon and nitrogen, and these could be picked up (remember we're viewing the future through rose-colored glasses) in virtually limitless quantities from the atmosphere of the gas giants, particularly Jupiter.

If we imagine 100,000 asteroids, all more or less hollowed out, we could end with a living space of 200,000,-000,000 square miles of 55,000 USA's. This would be more than 150 times as much area as was available on all the surfaces of the solar system (excluding the gas giants, but even including the asteroids).

Suppose the levels within an asteroid could be as densely populated as the United States today. We might then average 100,000,000 as the population of an asteroid, and the total population of all the asteroids would come to 10,000,000,000,000 (ten trillion).

The question is whether such a population can be supported. One can visualize each asteroid a self-sufficient unit, with all matter vigorously and efficiently cycled. (This, indeed, was the background of the story from which I quoted earlier.)

The bottleneck is bound to be the energy supply, since energy is the one thing consumed despite the efficiency with which all else is cycled.

At the present moment, virtually all our energy supply is

derived from the sun. (Exceptions are nuclear energy, of course, and energy drawn from tides or hot springs.) The utilization of solar energy, almost entirely by way of the green plant, is not efficient, since the green plant makes use of only 2 per cent or so of all the solar energy that falls upon the earth. The unutilized 98 per cent is not the major loss, however.

Solar radiation streams out in all directions from the sun, and when it reaches the earth's orbit, it has spread out over a sphere 93,000,000 miles in radius. The surface area of such a sphere is 110,000,000,000,000,000 (a hundred and ten quadrillion) square miles, while the cross-section area presented by the earth is only 50,000,000 square miles.

The fraction of solar radiation stopped by the earth is therefore 50,000,000/110,000,000,000,000,000, or just about 1/2,000,000,000 (one two-billionth).

If all the solar radiation could be trapped and utilized with no greater efficiency that it is now on earth, then the population supportable (assuming energy to be the bottleneck) would mount to two billion times the population of the earth or about 6,000,000,000,000,000,000 (six quintillion).

To be sure, the energy requirement per individual is bound to increase, but then efficiency of utilization of solar energy may increase also and, for that matter, energy can be rationed. Let's keep the six quintillion figure as a talking point.

To utilize all of solar radiation, power stations would be set up in space in staggered orbits at all inclinations to the ecliptic. As more and more energy was required, the station would present larger surfaces, or there would be more of them, until eventually the entire sun would be encased. Every bit of the radiation would strike one or another of the stations before it had a chance to escape from the solar system.

This would create an interesting effect to any intelligent being studying the sun from another star. The sun's visible light would, over a very short period, astronomically speaking, blank out. Radiation wouldn't cease altogether, but it

222

would be degraded. The sun would begin to radiate only in the infrared.

Perhaps this always happens when an intelligent race becomes intelligent enough, and we ought to keep half an eye peeled out for any star that disappears without going through the supernova stage—any that just blanks out.

Who knows?

An even more grisly thought can be expounded. From an energy consideration, I said that a human population of six quintillion might be possible.

On the other hand, the total population of the asteroids, at an American population density, was calculated at a mere ten trillion. Population could still increase 600,000-fold, but where would they find the room?

An increase in the density of the population might seem undesirable and, instead, the men of the asteroids might cast envious eyes on other worlds. Suppose they considered a satellite like Saturn's Phoebe, with its estimated diameter of 200 miles. It could be broken up into about two hundred small asteroids with a diameter of 35 miles each. Instead of one satellite with a surface area of 120,000 square miles, there would be numerous asteroids with a total internal area of 400,000,000 square miles.

The gain might not be great with Phoebe, for considerable hollowing out might be carried on upon that satellite even while it was intact. Still, what about the moon, where hollowing would have to be confined to the outermost skin?

It has a greater mass than all the asteroids put together, and if it were broken up, it would form 200,000 asteroids of 35-mile diameter. At a stroke, the seating capacity, so to speak, of the human race would be tripled.

One can envisage a future in which, one by one, the worlds of the solar system will be broken into fragments for the use of mankind.

But, of course, earth would be in a special class. It would be the original home of the human race, and sentiment might keep it intact.

Once all the bodies of the solar system, except for the gas giants and earth, are broken up, the total number of asteroids would be increased roughly ten-million-fold, and the total human population can then reach the maximum that the energy supply will allow.

But, and here is the crucial point, Pluto may offer difficulties. For one thing, we aren't too certain of its nature. Perhaps its makeup is such that it isn't suitable for breaking up into asteroids. Then, too, it is quite distant. Is it possible that it is too far away for energy to be transmitted efficiently from the solar stations to all the millions of asteroids that can be created from Pluto, out four billion miles from the sun?

If Pluto is ignored, then there is only one way in which mankind can reach its full potential, and that would be to use the earth.

I can see a long drawn-out campaign between the Traditionalists and the Progressives. The former would demand that the earth be kept as a museum of the past and would point out that it was not important to reach full potential population, that a few trillion more or less people didn't matter.

The Progressives would insist that the earth was made for man and not vice versa, that mankind had a right to proliferate to the maximum, and that in any case, the earth was in complete darkness because the solar stations between itself and the sun soaked up virtually all radiation, so that it could scarcely serve as a realistic museum of the past.

I have a feeling the Progressives would, in the end, win, and I pull down the curtain as the advancing workfleet, complete with force beams, prepares to make the preliminary incision that will allow the earth's internal heat to blow it apart as the first step in asteroid formation.